IONIZING RADIATION AND POLYMERS

PLASTICS DESIGN LIBRARY (PDL)

PDL HANDBOOK SERIES

Series Editor: Sina Ebnesajjad, PhD
President, FluoroConsultants Group, LLC
Chadds Ford, PA, USA
www.FluoroConsultants.com

The **PDL Handbook Series** is aimed at a wide range of engineers and other professionals working in the plastics industry, and related sectors using plastics and adhesives. PDL is a series of data books, reference works and practical guides covering plastics engineering, applications, processing, and manufacturing, and applied aspects of polymer science, elastomers and adhesives.

Recent titles in the series
Brandau, *Bottles, Preforms and Closures*, Second Edition
ISBN: 9781437735260

Brandau, *Stretch Blow Molding*, Second Edition
ISBN: 9781437735277

Chandrasekaran, *Rubber Seals for Fluid and Hydraulic Systems*
ISBN: 9780815520757

Ebnesajjad, *Handbook of Adhesives and Surface Preparation*
ISBN: 9781437744613

Ebnesajjad, *Handbook of Biopolymers and Biodegradable Plastics*
ISBN: 9781455728343

Grot, *Fluorinated Ionomers*, Second Edition
ISBN: 9781437744576

Kutz, *Applied Plastics Engineering Handbook*
ISBN: 9781437735147

Kutz, *PEEK Biomaterials Handbook*
ISBN: 9781437744637

McKeen, *Permeability Properties of Plastics and Elastomers*, Third edition
ISBN: 9781437734690

McKeen, *The Effect of Sterilization Methods on Plastics and Elastomers*, Third Edition

McKeen, *The Effect of Sterilization on Plastics and Elastomers*
ISBN: 9781455725984

Sastri, *Plastics in Medical Devices*
ISBN: 9780815520276

Sin, Rahmat & Rahman, *Polylactic Acid*
ISBN: 9781437744590

Tolinski, *Additives for Polyolefins*
ISBN: 9780815520511

Wagner, *Multilayer Flexible Packaging*
ISBN: 9780815520214

Woishnis & Ebnesajjad, *Chemical Resistance of Thermoplastics* (2 volume set)
ISBN: 9781455778966

Woishnis & Ebnesajjad, *Chemical Resistance of Specialty Thermoplastics: Chemical Resistance*, Volume 3
ISBN: 9781455731107

To submit a new book proposal for the series, please contact
Sina Ebnesajjad, Series Editor
sina@FluoroConsultants.com
or
Matthew Deans, Senior Publisher
m.deans@elsevier.com

IONIZING RADIATION AND POLYMERS

Principles, Technology, and Applications

By

Jiri George Drobny

Drobny Polymer Associates

ELSEVIER

AMSTERDAM • BOSTON • HEIDELBERG • LONDON
NEW YORK • OXFORD • PARIS • SAN DIEGO
SAN FRANCISCO • SINGAPORE • SYDNEY • TOKYO
William Andrew is an imprint of Elsevier

William Andrew is an imprint of Elsevier
The Boulevard, Langford Lane, Kidlington, Oxford, OX5 1GB
225 Wyman Street, Waltham, MA 02451, USA

First published 2013

Notices

Knowledge and best practice in this field are constantly changing. As new research and experience broaden our understanding, changes in research methods, professional practices, or medical treatment may become necessary.

Practitioners and researchers must always rely on their own experience and knowledge in evaluating and using any information, methods, compounds, or experiments described herein. In using such information or methods they should be mindful of their own safety and the safety of others, including parties for whom they have a professional responsibility.

To the fullest extent of the law, neither the Publisher nor the authors, contributors, or editors, assume any liability for any injury and/or damage to persons or property as a matter of products liability, negligence or otherwise, or from any use or operation of any methods, products, instructions, or ideas contained in the material herein.

British Library Cataloguing-in-Publication Data
A catalogue record for this book is available from the British Library

Library of Congress Cataloging-in-Publication Data
A catalog record for this book is available from the Library of Congress

ISBN: 978-1-4557-7881-2

For information on all William Andrew publications
visit our website at store.elsevier.com

To the memory of my parents and my sister Marta

Contents

Preface

Marshall R. Cleland, well-known scientist, engineer, inventor, and industrialist defines radiation processing as "the treatment of products and materials with ionizing radiation to change their physical, chemical, or biological properties, to increase their usefulness and value, or to reduce their impact on the environment."

When applying this definition to polymers, we can see that ionizing radiation is a well-established and economical commercial method of precisely modifying the properties of bulk polymeric systems and formed components made from them. The chemical reactions of cross-linking, chain scission, oxidation, grafting, and long chain branching of polymers initiated by radiation have found many useful applications in a great variety of plastic, elastomeric, and composite materials.

Important characteristics of polymeric materials, such as mechanical properties, thermal stability, melt flow, processability, chemical resistance, and surface properties can be significantly improved or modified by ionizing radiation.

Processing of polymers and various polymeric systems by radiation represents an important economical and environmental advantage over a number of established conventional processes which are using mainly heating and often complex formulations containing additives with negative environmental impact. Low pollution and low energy consumption of the ionizing radiation equipment and processes used in radiation processing of polymers with the potential recycling of the materials makes it a green technology with an exceptional attractiveness in the current efforts in sustainability. Radiation curing of coatings and inks can be solvent free or can use significantly reduced volatile organic compounds.

This book covers the current technology based on ionizing radiation, i.e., gamma ray, electron beam, and X-ray radiations. Its main goal is to provide a practical description and explanation of these processes with only essential theory necessary for readers to understand most of the material presented.

Chapter 1 (Introduction) describes the types of radiations in general and ionizing radiation in particular and some pertinent scientific principles involved. Chapter 2 covers the fundamentals of radiation chemistry. Chapter 3 deals with the current types of equipment, designs, and typical uses with particular attention given to electron beam because this type of machinery is the most widely used in industrial practice involving polymers and polymeric systems. Chapter 4 describes the electron beam processes, while Chapter 5 describes the electron beam processing of commercial polymeric materials. These two chapters also concentrate mainly on electron beam for the same reason as for Chapter 3. Chapter 6 summarizes the industrial applications of ionizing radiation. Radiation-resistant polymers and their uses are discussed in Chapter 7. Measuring methods for ionizing radiation are described in Chapter 8, and health and safety considerations in Chapter 9. Current state of technology and trends are summarized in Chapter 10. Current known equipment manufacturers are listed in Appendix I, and standards and specifications applicable to ionizing radiation in Appendix II, units used in ionizing radiations and conversions are summarized in Appendix III, and equations used in the text are reviewed in Appendix IV. Appendix V lists the 12 principles of green chemistry. A rather comprehensive glossary of relevant terms is also included.

There were many colleagues, friends, and other people who contributed in many ways during the preparation of the manuscript: Marshall Cleland provided numerous resources and valuable advice; Tony Berejka reviewed parts of the manuscript and offered valuable comments and data. Other people helped by providing photographs, data, information, and recommendations for improvements of the manuscript, including John Chrusciel, Werner Haag, Karl Swanson, Terrance Thompson, Bengt Laurell, Phillipe Dethier, Ann Car, Jeremy Simon, Gary Cohen, and others. Without encouragement and support from the Elsevier team (Matthew Deans, Sina Ebnesajjad, Frank Hellwig and David Jackson) and without the help and cooperation from Lisa Jones and her production team this work would not have come to fruition.

Jiri George Drobny
Merrimack, New Hampshire and
Prague, Czech Republic
June 2012

Abbreviations

ACS	American Chemical Society
AECL	Atomic Energy of Canada, Ltd.
ALARA	as low as reasonably achievable
APP	atmospheric plasma processing
ASTM	American Society for Testing and Materials
BOPP	bias oriented polypropylene
BR	butadiene rubber, polybutadiene
BRC	biorenewable carbon contents
CATV	community accessible television
CIIR	chlorinated isobutylene–isoprene rubber (chlorobutyl)
CMC	carboxymethylated cellulose
COPA	polyamide thermoplastic elastomer (US designation)
COPE	copolyester–ether block copolymer
CR	polychloroprene rubber
CSM	cure site monomer
CTA	cellulose triacetate
CV	continuous vulcanization
DCE	dichloroethane
DMSO	dimethylsulfoxide
DPGDA	di(propyleneglycol) diacrylate
DTMPTA	di(trimethylolpropane) tetraacrylate
DVB	divinylbenzene
EB	electron beam
ECTFE	copolymer of ethylene and chlorotrifluoroethylene
EPC	Easy Processing Channel Black
EPDM	ethylene–propylene–diene monomer rubber
EPM	ethylene–propylene rubber (also abbreviated as EPR)
EPR	electron paramagnetic resonance (also)
ETFE	copolymer of ethylene and tetrafluoroethylene
FEP	copolymer of hexafluoropropylene and tetrafluoroethylene
FFKM	perfluoroelastomer (ASTM designation)
FKM	fluorocarbon elastomer (ASTM designation)
FMQ	fluorosilicone rubber (with methyl phenyl substituents)
FPM	fluorocarbon elastomer (ISO designation)

FTIR	Fourier transform infrared (analysis)
FVMQ	fluorosilicone rubber (with methyl and vinyl substituents)
G(S)	chain scission G-value, number of chains scissions occurring per 100 eV of absorbed energy
G(X)	cross-linking G-value, number of cross-links occurring per 100 eV of absorbed energy
GMA	glycidyl methacrylate
GPTA	glyceryl propoxytriacrylate
HAF	high abrasion furnace black
HCl	hydrochloric acid
HDDA	1,6-hexanediol diacrylate
HDPE	high-density polyethylene
HEMA	hydroxymethyl methacrylate
HF	hydrofluoride
HFP	hexafluoropropylene
IAEA	International Atomic Energy Agency
IcRP	International Commision on Radiological Protection
IPN	interpenetrating polymer network
IR	cis-1.4 isoprene rubber, polyisoprene rubber (also infrared)
ISO	International Standard Organization
LDPE	low-density polyethylene
LET	linear energy transfer
LLDPE	linear low-density polyethylene
M_c	molecular weight between cross-links
M_w	weigh average molecular weight
MA	maleic anhydride
MFA	1. multifunctional acrylate 2. copolymer of tetrafluoroethylene and perfluoromethyl vinyl ether
MPBM	N-N′ - (m-phenylene)-bismaleimide
MSDS	Material Safety Data Sheets
MW	molecular weight
N	cross-link density
NBR	acrylonitrile–butadiene rubber (“nitrile rubber”)
NIST	National Institute of Science and Technology
nm	nanometer
NMR	nuclear magnetic resonance
NR	natural rubber
NVP	N-vinl pyrrolidone
OIT	oxygen (or oxidative) induction time
OSHA	Occupational Safety and Health Administration
OSL	Optically stimulated luminiscence

PAI	polyamide–imide
PCTFE	poly(chlorotrifluoroethylene)
PDMS	polydimethylsiloxane
PE	polyethylene
PEEK	poly(ether ether ketone)
PEG	poly(ethylene glycol)
PEI	poly(ether imide)
PEO	poly(ethylene oxide)
PES	poly(ether sulfone)
PET	poly(ethylene terephthalate)
PETA	pentaerythritol triacrylate
PEX	cross-linked polyethylene pipes
PFA	copolymer of tetrafluoroethylene and perfluoro (propylvinyl ether)
PI	polyimide
PMMA	poly(methyl methacrylate)
PMVE	perfluoro(methylvinyl ether)
PP	polypropylene
PPS	poly(phenylene sulfide)
PPVE	perfluoro(propylvinyl ether)
PSA	pressure sensitive adhesives
PSU	polysulfone
PTFE	poly(tetrafluoroethylene)
PVA	poly(vinyl alcohol)
PVC	poly(vinyl chloride)
PVDC	poly(vinylidene chloride)
PVDF	poly(vinylidene fluoride)
PVF	poly(vinyl fluoride)
R	gas constant
rad	radiation-absorbed dose
REACH	Registration, Evaluation, and Restriction of Chemical Substances
rem	roentgen-equivalent man
RF	radio frequency
RRC	radiation rapid curing
RTM	resin transfer molding
RV	radiation vulcanized
SAL	sterilization assurance level
SBC	styrenic block copolymer
SBR	styrene–butadiene rubber
SBS	styrene–butadiene–styrene (block copolymer)

SEBS	styrene–ethylene–butylene–styrene (block copolymer)
SEEPS	triblock styrenic copolymer
SEPS	styrene–ethylene–propylene–styrene (block copolymer)
SiBS	styrene–isobutylene–styrene (block copolymer)
SIBS	styrene–isoprene–butadiene–styrene (block copolymer)
SIS	styrene–isoprene–styrene (block copolymer)
STP	standard temperature and pressure
T	absolute temperature
T_g	glass transition temperature
T_m	crystalline melting temperature
TAC	triallyl cyanurate
TAIC	triallyl isocyanurate
TFE	tetrafluoroethylene
TLD	Thermoluminiscent dosimeter
TMET	trimethylolethane trimethacrylate
TMP(EO)TA	trimethylolpropane ethoxytriacrylate
TMP(PO)TA	trimethylolpropane propoxytriacrylate
TMPTA	trimethylolpropane triacrylate
TMPTMA	trimethylolpropane trimethacrylate
TPE	thermoplastic elastomer
TPGDA	tripropyleneglycol diacrylate
TPO	thermoplastic polyolefin (thermoplastic elastomer)
TPU	thermoplastic polyurethane (thermoplastic elastomer)
TPV	thermoplastic rubber vulcanizate (thermoplastic elastomer)
TSCA	Toxic Substances Control Act
UHMWPE	ultrahigh-molecular-weight polyethylene
UV	ultraviolet
VDF	vinylidene fluoride
VHF	very high frequency
VOC	volatile organic compound
W & C	wire and cable
WFRP	wood fiber reinforced plastics
WPC	wood–plastic composite
XLPE	cross-linked polyethylene
XPE	cross-linked polyethylene

1 Introduction

OUTLINE

Radiant energy is one of the most abundant forms of energy available to humankind. Nature provides sunlight, the type of radiation essential for many forms of life and growth. Some natural substances, such as radioactive elements, generate the kind of radiation that can be destructive to life, but when harnessed, it can be very useful for medical and industrial applications.

1.1 Radiation Types and Sources

Devices for the generation of radiant energy useful in a great variety of scientific, industrial, and medical applications have been invented and perfected. *Cathode ray* tubes emit impulses that activate some screens of televisions and computer monitors. *X-rays* are used not only as diagnostic tools in medicine but also as an analytical tool in the inspection of manufactured products, such as tires and other composite structures as well as in the manufacture of polymer-based composites. *Microwaves* are used not only in cooking but also in heating of certain materials and a variety of electronic applications including ultraviolet (UV) curing lamps. *Infrared (IR) radiation* is used in heating, analytical chemistry, and a variety of electronic devices. *Ion beams* have been used extensively in commercial ion implantation in the production of semiconductor devices and also for surface hardening of metals, although their use in the processing of polymeric materials is limited to only specialized applications, such as treatment of polymer surfaces and thin films because they have extremely low penetration depths [1].

Drobny: Ionizing Radiation and Polymers.
DOI: http://dx.doi.org/10.1016/B978-1-4557-7881-2.00001-8

Table 1.1 Frequency and Wavelength of Various Types of Electromagnetic Radiation

Radiation	Wavelength (μm)	Frequency (Hz)
Infrared	$1-10^{2}$	$10^{15}-10^{12}$
Ultraviolet	$10^{2}-1$	$10^{17}-10^{15}$
Microwave	$10^{3}-10^{5}$	$10^{12}-10^{10}$
X-rays soft	$10^{-2}-10^{-3}$	$10^{17}-10^{16}$
X-rays hard	$10^{-4}-10^{-3}$	$10^{19}-10^{17}$
Electron beam	$10^{-7}-10^{-5}$	$10^{21}-10^{18}$
Gamma rays	$10^{-6}-10^{-5}$	$10^{-20}-10^{18}$

Laser beam radiation has found a wide use in medical and military applications, and in numerous industrial applications.

*UV and electron beam (EB) radiation*s are classified as *electromagnetic radiation* along with IR, visible, gamma, microwave, and laser beam radiations. The differences between various electromagnetic radiations are given in Table 1.1.

The widely used industrial processes involving man-made electromagnetic radiations, namely UV and EB types, depend essentially on two electrically generated sources of radiation: photons from high-intensity UV lamps and accelerated electrons, respectively. The difference between those two is that accelerated electrons that penetrate matter are stopped only by mass, whereas high-intensity UV light affects only the surface. Both EB and UV radiations represent a clean and efficient use of electric energy. In the following chapters, we will limit our attention mainly to *ionizing radiation*, which includes EB, gamma, and X-ray radiations, with a particular stress on the EB equipment, processes, and applications.

1.2 Ionizing Radiations

Ionizing radiation can modify physical, chemical, and biological properties of the irradiated materials. At present, the principal industrial applications of radiation are sterilization of health-care products including pharmaceuticals, irradiation of food and agriculture products (for various end objectives, such as disinfestation, shelf life extension, sprout inhibition, pest control, and sterilization), and materials modification (such as polymerization, polymer cross-linking, and gemstone colorization).

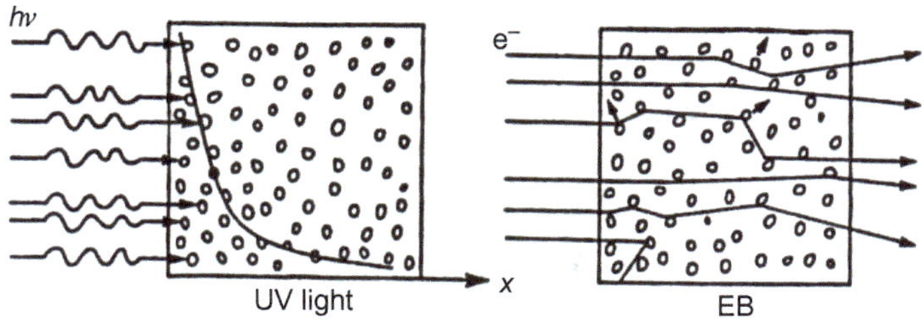

Figure 1.1 Comparison of UV and EB Radiations (Equal Substrate Thickness).

The main three types of ionizing radiation include high-energy electrons (EB), gamma rays (γ-rays), and X-rays. These are capable of not only converting monomeric and oligomeric liquids into solids but also producing major changes in properties of solid polymers. Also, in comparison to UV and visible radiations, they can penetrate considerably deeper into the material as illustrated in Figure 1.1.

1.2.1 Gamma Rays

γ-Rays represent electromagnetic radiation emitted from excited atomic nuclei of unstable atoms, the so-called radionuclides, as an integral part of the process whereby the nucleus rearranges itself into a state of lower excitation (i.e., lower energy content). A γ-ray is a packet of electromagnetic energy—photons. These photons are the most energetic ones in the electromagnetic spectrum. Essentially, they are emitted by *radioactive decay* and have energies in the range from 10^4 to 10^7 eV. All γ-rays emitted from a given *radioactive isotope* have the same energy. γ-Rays penetrate matter farther than *beta* or *alpha* particles, producing ionization (electron disruption) in their path. In living cells, these disruptions result in damage to the DNA and other cellular structures eventually causing the death of the organism or render it incapable to reproduce. γ-Rays do not create residuals or impart radioactivity in the materials exposed to them. In that they are similar to X-rays (see Section 1.2.2). They ionize matter by three main processes: the photoelectric effect, Compton scattering, and pair production. In the wide energy range of 100 keV to 1 MeV, Compton scattering is the main absorption mechanism, in which an incident γ-photon loses enough energy to eject an electron in an atom of the irradiated matter, and the remainder of its energy is emitted as a new γ-photon with lower energy. γ-Irradiation using ^{60}Co source has a low-dose rate or dose absorbed by the matter per unit time (on the order of 10^{-3} kGy/s). The dose rate of γ-rays is much lower than that of an EB. The comparison of the sources of ionizing radiation is given in Table 1.2.

Compared to EB, gamma irradiation has a much higher penetration, which is an advantage when irradiating products with large volume. However, the absorbed dose decreases exponentially with the increasing depth of penetration following the Lambert–Beer law:

$$I_t = I_0 \, e^{-at} \tag{1.1}$$

where I_t is the intensity of the radiation after passing through the thickness t, I_0 is the initial intensity, and a is the coefficient of linear absorptivity. The attenuation reduces the dose uniformity across the matter.

The most widely used radioactive isotopes used in medical and industrial applications are cobalt 60 (^{60}Co), cesium 137 (^{137}Cs), and iridium 192 (^{192}Ir). The half-life of ^{60}Co is 5.3 years, that of ^{137}Cs is 30 years, and the half-life of ^{192}Ir is 74 days [1]. (The *half-life* is defined as the period of time in which half the nuclei of a species of radioactive substance would decay.) When used for irradiation, the isotope is generally in the form of a pellet size 1.5×1.5 mm^2 loaded into a stainless steel capsule and sealed or in the form of rods or "pencils."

Unlike EB or X-rays, γ-rays cannot be turned off. Once radioactive decay starts, it continues until all the atoms have reached a stable state. The radioisotope can only be shielded to prevent exposure to the radiation. The most common applications of γ-rays are sterilization of single-use medical supplies, elimination of organisms from pharmaceuticals, microbial reduction in and on consumer products, cancer treatment, and processing of polymers (cross-linking, polymerization, degradation, etc.). It should be noted that the products that were irradiated by γ-rays do not become radioactive and thus can be handled normally.

Strength (or power) of a γ-radiation source is referred to as *radioactivity*. It is defined as spontaneous changes in a nucleus accompanied by the emission of energy from the nucleus as a radiation. Units of radioactivity are *curie* (Ci) or *bequerel* (Bq). Radioactivity is defined as the number of disintegrations (dis) of radioactive nuclides per second. Bequerel is the SI unit of radioactivity, which is 1 dis/s. However, this is a very small amount of radioactivity, and traditionally it is measured in units of curie. 1 curie $= 3.7 \times 10^{10}$ Bq. For a γ-radiator, source power may be calculated from the source activity, such that 10^6 curie is approximately 15 kW of power.

1.2.2 X-Rays

X-radiation (also called Röntgen radiation) is a form of electromagnetic radiation. X-rays have a wavelength in the range from 10 to 0.01 nm,

Table 1.2 Comparison of Sources of Ionizing Radiation

Characteristic	Source of Radiation		
	Gamma Rays	**Electron Beams**	**X-Rays**
Power source	Radioactive isotope[a]	Electricity	Electricity
Power activity	Half-life 5.27 years	Electrical on–off	Electrical on–off
Properties	Photons (1.25 MeV) $\lambda = 1.0 \times 10^{-3}$ nm	Electrons Mass $= 9.1 \times 10^{-31}$ kg	Photons $\lambda = 4.1 \times 10^{-3}$ nm
Charge	None	1.60×10^{-9} C	None
Equipment	Easy to operate and maintain	Complicated to operate and maintain	Complicated to operate and maintain
Emission	Isotropic, cannot be controlled	Unidirectional (can be scanned and bent by magnets)	Forward peaked
Penetration	Exponential attenuation	Finite range, depending on energy	Exponential attenuation
Source attenuation	Continuous attenuation requires regular addition of source	No attenuation	No attenuation
Shielding	Continuous operation requires more shielding	Can be switched on and off, less demanding on shielding	Can be switched on and off, less demanding on shielding
Dose rate	10 kGy/h; 2.8×10^{-3} kGy/s	360,000 kGy/h; 100 kGy/s	960 kGy/h; 0.27 kGy/s

[a] *Mainly* ^{60}*Co.*

corresponding to frequencies ranging from 3×10^{16} to 3×10^{19} and energies in the range from 120 eV to 120 keV. They are considerably shorter in wavelength than the UV rays. Their names come from their discoverer Wilhelm Conrad Röntgen, who called them X-rays to signify an unknown type of radiation [2].

X-rays span three decades in wavelength, frequency, and energy. From about 0.12 to 12 keV, they are classified as soft X-rays, and from 12 to 120 keV, as hard X-rays, due to their penetrating abilities. Since X-rays are a form of ionizing radiation, they can be dangerous to living organisms.

There are two different atomic processes that can produce X-ray photons. One process produces *bremsstrahlung* (from German, meaning "breaking radiation") and the other produces *K-shell* or *characteristic* emission. Both processes involve a change in the energy state of electrons. X-rays are generated when an electron is accelerated and then made to rapidly decelerate usually due to interaction with other atomic particles. In an X-ray system, a large amount of electric current is passed through a tungsten filament, which heats the filament to several thousand degrees centigrade to create a source of free electrons. A large potential is established between the filament (the cathode) and a target (the anode). The two electrodes are in a vacuum. The electrical potential between the cathode and anode pulls electrons from the cathode and accelerates them, as they are attracted toward the anode, which is usually made of tungsten (see Section 1.2.3). X-rays are generated when free electrons give up the energy as they interact with the orbital electrons or nucleus of an atom. The interaction of the electrons in the target results in the emission of a continuous radiation spectrum and also characteristic X-rays from the target material. Thus the difference between γ-rays and X-rays is that the γ-rays originate in the nucleus and X-rays originate in the electrons outside the nucleus (surrounding it) or are produced in an X-ray generator [3–6].

Despite the high-power and high-dose rate, EB that is used for the production of X-rays has a serious limitation in the penetration of thick products. However, if the EB is converted into X-rays, it overcomes the problem of low penetration. The conversion process to X-rays has a low-efficiency, and X-ray irradiation has been made commercially feasible only with the development of high-energy and high-beam power electron accelerators.

X-rays are primarily used for diagnostic radiography in medicine [7] and crystallography. Other notable uses are X-ray microscopic analysis, X-ray fluorescence as an analytical method, and industrial radiography for the inspection of industrial parts, such as tires and inspection of welds [8]. Recent reports and patents cover the use of X-rays in processing of variety of parts made from polymers, particularly in the form of advanced fiber-reinforced composites [9–13]. Similarly to γ-rays, the products that were irradiated by X-rays do not become radioactive.

1.2.3 EB Radiation

In principle, fast electrons are generated in a high vacuum (typically 10^{-6} Torr) by a heated cathode. The electrons emitted from the cathode are then accelerated in an electrostatic field applied between cathode and anode. The acceleration takes place from the cathode that is on a negative high-voltage potential to the grounded vessel as anode. The accelerated electrons may sometimes be focused by an optical system to the window plane of the accelerator [14].

The energy gain of the electrons is proportional to the *accelerating voltage* and is expressed in *electron volts* (eV), which represent the energy gained by a particle of unit charge by passing the potential difference of 1 V. The electrons leave the vacuum chamber only if their energy is high enough to penetrate the 15–20 μm thick titanium window of the accelerator. Details about the equipment used to produce EB radiation are given in Chapter 3.

When an EB enters a material (this includes the accelerator exit window, the air gap, and the material being irradiated), the energy of the accelerated electrons is greatly altered. They lose their energy and slow down almost continuously as a result of a large number of interactions each with only small energy loss.

Electrons, as do any other charged particles, transfer their energy to the material through which they pass in two types of interaction:

1. In *collisions with electrons* of an atom resulting in material ionization and excitation.
2. In *interaction with atomic nuclei* leading to the emission of X-ray photons, i.e., bremsstrahlung. Bremsstrahlung is essentially an electromagnetic radiation emitted when a charged particle slows down due to such interaction as pointed out earlier.

1.3 Ion Beam Radiation

An ion beam is a type of particle beam consisting of ions. High-energy ion beams are produced like EBs by particle acceleration, with the difference that a cyclotron, another particle accelerator (Figure 1.2), is used. Irradiation effects of an ion beam are different from ionizing radiation due to the differences in the *linear energy transfer* (LET), the average energy deposited in the material used by a projectile particle along its path. In general, the LET of an ion beam is larger than that of EB, depending on the particle mass and energy. Ion beam radiation has been used extensively in commercial applications for ion implantation in the production of semiconductor devices and surface hardening of metals. However, over the past decade, ion beams

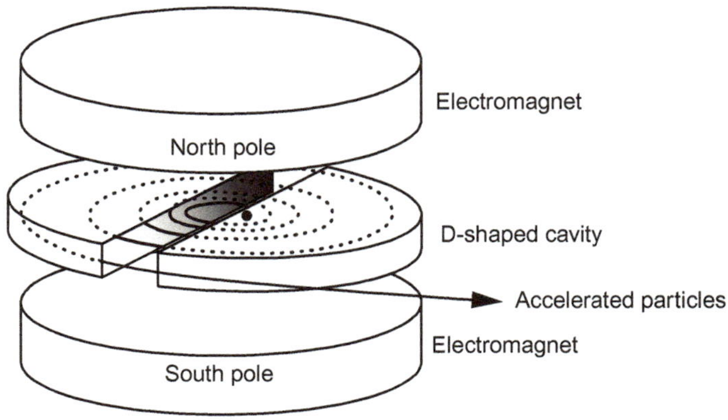

Figure 1.2 Schematic of a Cyclotron.

have been adapted for commercial use for the processes involving polymers. Initially, their applications had included treatment of polymer surfaces and thin films because of their very low depth penetration [15], but recent reports in literature pertain to their use in modifications of mechanical properties of carbon-fiber-reinforced plastics [16,17], in surface modification of polytetrafluoroethylene (PTFE) [18], and in the production of fuel cells [19].

1.4 Laser Beam Radiation

A laser is a device that emits light (electromagnetic radiation) through a process of optical amplification based on the stimulated emission of photons. The term "laser" originated as an acronym for *Light Amplification by Stimulated Emission of Radiation.* The emitted laser light is notable for its high degree of spatial and temporal coherence, unattainable using other technologies.

Besides a wide use in medical and military applications, lasers are used in industrial applications for cutting, welding, material heat treatment, marking parts, and noncontact measurement of parts [20].

References

[1] NDT Resource Center, <www.ndt-ed.org>; November 5, 2009.

[2] Kevles BH. Naked to the bone medical imaging in the twentieth century. Camden, NJ: Rutgers University Press; 1996. p. 19–22.

[3] Dendy PP, Heaton B. Physics for diagnostic radiology. Boca Raton, FL: CRC Press; 1999. p. 12.

[4] Feynman R, Leighton R, Sands M. Feynman lectures on physics, vol. 1. Boston: Addison-Wesley; 1963. p. 2–5.
[5] L'Annunziata M, Baradei M. Handbook of radioactivity analysis. Waltham, MA: Academic Press; 2003. p. 58.
[6] Grupen C, Cowan G, Eidelman SD, Stroh T. Astroparticle physics. Heildelberg: Springer; 2005. p. 109.
[7] Smith MA, Lundahl B, Strain P. Med Device Technol 2003;16(3):16–8.
[8] Drobny JG. Radiation technology for polymers. Boca Raton, FL: CRC Press; 2010. p. 16.
[9] Ramos MA, Catalao MM, Schacht E, Mondalaers W, Gil MH. Macromol Chem Phys 2002;203(10–11):1370–6.
[10] Wang C-H, et al. J Phys D 2008;41(19):8 [paper 195301].
[11] Sanders CB, et al. Radiat Phys Chem 1995;46(4–6):991–4.
[12] Berejka AJ. Electron beam curing of composites: opportunities and challenges. RadTech Rep 2002;16(2):33.
[13] Galloway RA, Berejka AJ, Gregoire O, Clelland MR. Processes for chemically reactive materials with X-rays. U.S. patent application 20080196829; 2008.
[14] Eckstrom DJ, et al. J Appl Phys 1988;64:1691.
[15] Clough RL. Nucl Instrum Methods Phys Res B 2001;185:11.
[16] Kudoh H, Sasuga T, Seguchi. High energy ion irradiation effects on mechanical properties of polymeric materials. Radiat Phys Chem 1996;48(5):545.
[17] Seguchi T, et al. Ion beam irradiation effect on polymers. LET dependence on the chemical reactions and change of mechanical properties. Nucl Instrum Methods Phys Res B 1999;151:154.
[18] Choi YJ, Kim MS, Noh I. Surface modification of a polytetrafluoroethylene film by cyclotron ion beams and its evaluation. Surf Coat Technol 2007;201:5724.
[19] Yamaki T, et al. Nano-structure controlled polymer electrolyte membrane for fuel cell applications prepared by ion beam irradiation. Proceedings of conference proton exchange fuel cells. Cancun, Mexico: Electrochemical Society, Inc; October 29–November 3, 2006.
[20] Taniguchi N, Ikeda M, Miyamoto I, Miyazaki T. Energy-beam processing of materials. Oxford: Clarendon Press; 1989.

Recommended Further Reading

Makuuchi K, Cheng S. Radiation processing of polymer materials and its industrial applications. Hoboken, NJ: John Wiley & Sons; 2012.

Industrial Radiation Processing with Electron Beams and X-rays. International Atomic Energy Agency. Vienna, Austria; 2011.<www.iaea.org>.

Gamma Irradiators for Radiation Processing. International Atomic Energy Agency. Vienna, Austria. <www.iaea.org>.

Drobny JG. Radiation technology for polymers. Boca Raton, FL: CRC Press; 2010.

L'Annunziata M, Baradei M. Handbook of radioactivity analysis. Waltham, MA: Academic Press; 2003.

Singh A, Silverman J, editors. Radiation processing of polymers. Munich: Hanser Publishers; 1992.

Taniguchi N, Ikeda M, Miyamoto I, Miyazaki T. Energy-beam processing of materials. Oxford: Clarendon Press; 1989.

Cuomo JJ, Rossnagel SM, Kaufman HR. Handbook of ion beam processing technology. Park Ridge, NJ: Noyes Publishing; 1989.

Feynman R, Leighton R, Sands M. Feynman lectures on physics, vol. 1. Boston: Addison-Wesley; 1963.

Charlesby A. Atomic radiation and polymers. Oxford, UK: Pergamon Press; 1960.

2 Fundamentals of Radiation Chemistry and Physics

OUTLINE

2.1 Introduction

The primary effect of any ionizing radiation is based on its ability to excite and ionize molecules, and this leads to the formation of free radicals, which then initiate reactions, such as polymerization and/or cross-linking or polymer degradation.

When the energy transferred to the irradiated material is higher than a particular orbital electron, the electron is ejected and the atom is ionized. However, if the energy is not high enough for ionization, the electron is raised to an upper energy level, resulting in excitation (Figure 2.1) [1, p. 1]. The ionizing potential for most molecules is less than 15 eV, while the energies of industrial irradiators are in the range from 100 keV to 10 MeV; thus, the ionization is the predominant process [1, p. 2].

In general, the changes in a material depend on the amount of *absorbed energy*, regardless of whether it was induced by a natural source or by naturally accelerated electrons. Both types of the commercially used radiations will decrease in intensity when passing through the matter. This happens because of scattering and the transfer of energy to the irradiated material.

The mean value of absorbed energy by the unit of mass of processed material is referred to as *absorbed dose* or simply *dose*. The unit of absorbed dose is 1 gray (Gy) equal to 1 J/kg. A more practical unit for the absorbed dose is 1 kGy = 1 J/g. The older unit, used officially until 1986 and still used

Drobny: Ionizing Radiation and Polymers.
DOI: http://dx.doi.org/10.1016/B978-1-4557-7881-2.00002-X

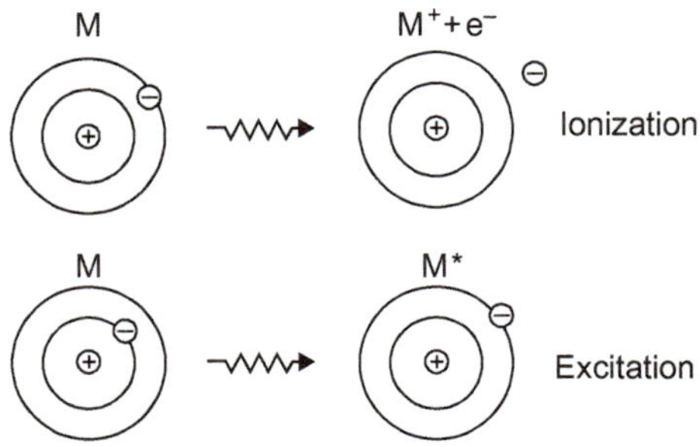

Figure 2.1 Ionization and Excitation. *(Reprinted with permission from John Wiley & Sons).*

Table 2.1 Dose Requirements for Various Radiation Effects

Radiation Effect	**Dose Requirement**
Radiography (film)	10^{-3}–10^{-2} Gy
Human lethal dose (LD50)	4–5 Gy
Sprout inhibition (potatoes, onions)	100–200 Gy
Potable water cleanup	250–500 Gy
Insect control (grains and fruits)	250–500 Gy
Wastewater disinfection	0.5–1 kGy
Fungi and mold control	1–3 kGy
Food spoilage bacteria control	1–3 kGy
Municipal sludge disinfection	3–10 kGy
Bacterial spore sterilization	10–30 kGy
Virus particle sterilization	10–30 kGy
Smoke scrubbing (SO_2 and NO_x)	10–30 kGy
Polymerization of monomers	10–30 kGy
Modification of polymers	10–50 kGy
Degradation of cellulosic materials	50–250 kGy
Degradation of scrap PTFE ("Teflon")	500–1500 kGy

occasionally in literature, is 1 megarad (Mrad), which is equivalent to 10 kGy. Dose requirements for various radiation effects are listed in Table 2.1.

The γ-rays generated by the radioactive decay interact with the irradiated material through secondary electrons. The typical energy of γ-rays is a few hundred eV [1, p. 2] higher than UV light and slightly higher than X-rays. γ-Rays ionize matter by three main processes: (1) the photoelectric effect, (2) Compton scattering, and (3) pair production. In the wide energy range of 100 keV–1 MeV, Compton scattering is the main absorption mechanism, in which an incident γ-photon loses enough energy to eject an electron in an atom of the irradiated material and the remainder of its energy is emitted as a new γ-photon with lower energy [2].

EB (electron beam) irradiators produce an EB by energizing and accelerating a stream of electrons through an electromagnetic or electrostatic field. The strength of the EBs as radiation is controlled by accelerating voltage and beam current. The beam current determines the number of accelerated electrons: 1 ampere (A) of current has a flow of 6.3×10^{18} per second. The commonly used electron energy ranges from 100 keV to 10 MeV and the power in the beam ranges from 0.5 to 200 kW [1, p. 6]. In contrast to γ-rays, EBs generated from an accelerator are monoenergetic. EB irradiators have much higher power than γ-irradiators.

When an EB enters a material, the energy of the accelerated electrons is greatly altered. They lose their energy and slow down almost continuously as a result of a large number of interactions with only a small energy loss.

X-rays are essentially photons produced by *bremsstrahlung* emitted from a metallic target. *Bremsstrahlung* (from German, meaning "breaking radiation") is thus an electromagnetic radiation emitted when a charged particle changes its velocity due to such an interaction. The yield is determined by the atomic number and thickness of the target and current of the incident EB: the higher the atomic number of the target, the higher the X-ray intensity. The bremsstrahlung photons are not monoenergetic but have a distribution over a range of energies. The penetration of X-rays is similar to that of γ-rays, but dose distribution may be more complicated [1, p. 12].

2.2 Interaction of High-Energy Electrons with Organic Matter

When an EB enters a material (this includes the accelerator exit window, the air gap, and the material being irradiated), the energy of the accelerated electrons is greatly altered. They lose their energy and slow down almost continuously as a result of a large number of interactions, each with only a small energy loss. As do any other charged particles, electrons transfer their energy to the material through which they pass in two types of interactions:

1. In collision with electrons of an atom resulting in material ionization and excitation.
2. In interaction with atomic nuclei leading to the emission of X-ray photons.

Somewhat simplified, the process of interaction of high-energy electrons with organic matter can be divided into three primary events:

1. *Ionization*: Ionization takes place only when the transferred energy during the interaction is higher than the bonding energy of the bonding electron:

$$AB \rightarrow AB^{+} + e^{-}$$

At almost the same time, the ionized molecule dissociates into a free radical and a radical ion:

$$AB^{+} \rightarrow AB\bullet \quad + \quad \bullet B^{+}$$

2. *Excitation*: Excitation moves the electron from the ground state to the excited state:

$$AB \rightarrow AB^{*}$$

The excited molecule eventually dissociates into free radicals:

$$AB^{*} \rightarrow A\bullet \quad + \quad B\bullet$$

3. *Capture of electron*: This process is also called ionization. Electrons with still lower energy can be captured by molecules. The resulting ion can dissociate into a free radical and a radical ion:

$$AB + e^{-} \rightarrow AB^{-}$$

$$AB^{-} \rightarrow A\bullet \quad + \quad \bullet B^{-}$$

Besides these primary reactions, there are various secondary reactions, in which ions or excited molecules take parts. The final result of these three events is that, through the diverse primary and secondary fragmentations,

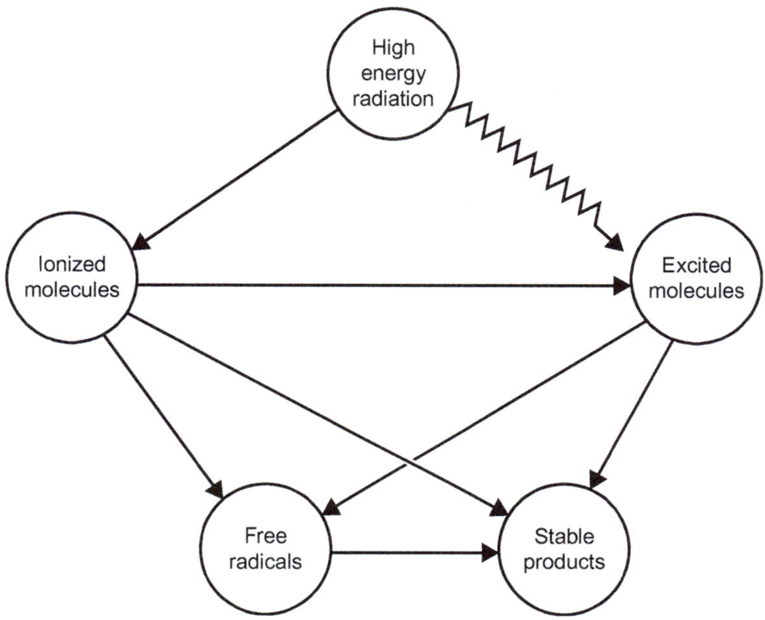

Figure 2.2 Radiation-Induced Events.

radicals are formed that can initiate a free-radical process leading to polymerization, cross-linking, backbone or side-chain scissions, structural rearrangements, etc. The complete cascade of reactions triggered by the primary excitation of molecules may take up to several seconds. The energy deposited does not always cause change in the precise position where it was originally deposited, and it can migrate and affect the product yield considerably. The radiation-induced events are depicted by a simplified schematic as shown in Figure 2.2.

Electrons that are capable of electronically exciting and ionizing organic molecules, such as acrylates and epoxides, must have energies in the range from 5 to 10 eV. Such electrons can be produced from fast electrons by energy degradation processes in solids, liquids, and gases. These secondary electrons show energy distribution with the maximum in the range from 50 to 100 eV. In contrast to fast electrons, exhibiting energies in the keV and MeV range, secondary electrons are capable of penetration in solids and liquids only a few nanometers. Consequently, they generate ions, radicals, and excited molecules in "droplets" along the paths of the fast electrons. Samuel and Magee [3] denoted such droplets containing several pairs of ions, radicals, and excited molecules as "spurs."

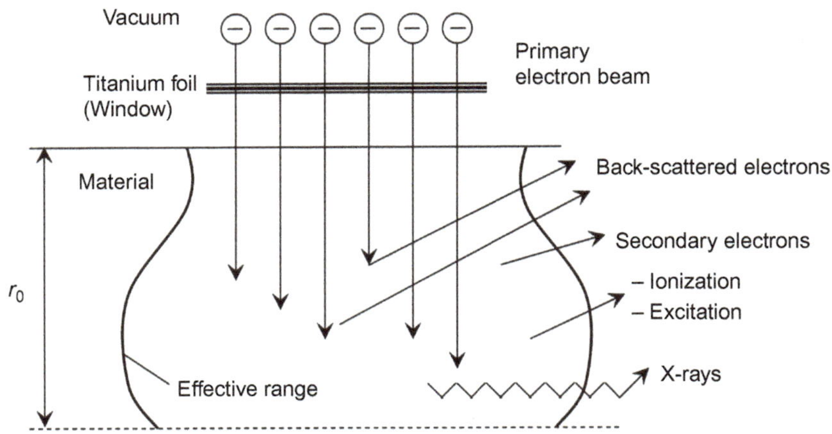

Figure 2.3 The Process of Generation of Reactive Species by High-Energy Electrons. (r_0 – penetration depth). *(Reprinted with permission from Vincentz Network.)*

2.2.1 Interaction of Ionic Radiation with Polymeric Materials

Chemical processing by EB involving polymerization and cross-linking of monomers and oligomers, and cross-linking, modifications, grafting, and degradation of polymers is induced by different reactive species formed initially by electrons in the spurs [4–6] as pointed out in the previous section. Figure 2.3 illustrates schematically the process of generation of reactive species [7]. Free radicals, which are important reactive species in polymer processing by ionizing radiation, are created either through scission of the main polymer chain or through the dissociation of the C–H side chain. Below are the main processes resulting from the interaction of ionizing radiation with polymers, oligomers, and monomers:

- *Cross-linking* is a process in which polymer chains are joined and a three-dimensional network is formed.
- *Long-chain branching* is a process in which polymer chains are joined but the network is not yet formed.
- *Chain scission* results in the reduction of molecular weight of the polymer, caused most frequently by oxidation or other forms of degradation. Oxidation and chain scission frequently occur simultaneously.
- *Polymerization* is initiated by the irradiation of monomers and/or oligomers. The combination of polymerization and cross-linking is referred to as *radiation curing*.

- *Grafting* is a chemical reaction proceeding by polymerization of a new monomer and attachment of the new chains to the main chain of the base polymer.

The mechanisms of the free-radical polymerization, cross-linking, and main chain scissions (polymer degradation) are given in Table 2.2.

Since different polymers respond differently to radiation, it is useful to quantify the response, namely in terms of cross-linking and chain scission. A parameter called the *G*-value is frequently used for this purpose. It represents the chemical yield of cross-links, scissions, double bonds, etc. of radiation by the number of molecules reacted per 100 eV (1.602×10^{-17} J) dissipated in the material, and in SI units it is expressed in micro-mols per joule of absorbed energy (μmols/J). The *G*-value for cross-linking is denoted as $G(X)$ and that for chain scission as $G(S)$. For example, the value of $G(X) = 4.5$ means that 4.5 cross-links are formed in the polymer per 100 eV under certain radiation conditions. In order to determine the number of cross-links or $G(X)$ and the number of scissions or $G(S)$, it is necessary to know the dose or dose rate and the time of exposure for these irradiation conditions. From the product yields, it is possible to estimate what ratio of monomer units is affected by irradiation [8].

For cross-linked materials, changes in cross-link density are reflected by the extent to which the material is swelled while being immersed in a compatible solvent. If cross-linking predominates, the cross-link density increases and the extent of swelling decreases. If chain scission predominates, the opposite occurs. An increased soluble fraction then reflects scission, whereas reduced soluble fraction reflects cross-linking.

For an uncross-linked material that undergoes predominantly cross-linking when exposed to ionizing radiation, solvent extraction experiments reveal that at certain absorbed dose (the "critical" dose), a percentage of the material is converted into insoluble gel. Beyond that point, the percentage of gel increases as a function of irradiation dose. In general, both the degrees of cross-linking and scissions can be determined from the soluble fraction by using the *Charlesby–Pinner equation* [9]:

$$s + s^{1/2} = p_0/q_0 + 10/q_0\, Du_1 \tag{2.1}$$

where s is the soluble fraction, p_0 is the density of scissions per unit dose, q_0 is the density of cross-links per unit dose, D is the absorbed dose in kGy, and u_1 is the number average molecular weight. Then the quantity p_0/q_0 represents the ratio of chain scissions to cross-links. The plot of $s + s^{1/2}$ versus $1/D$, a straight line, can be used to determine the p_0/q_0 ratio and from that the values of $G(S)$ and $G(X)$ can be calculated (Figure 2.4).

Table 2.2 Reactions Induced by Ionizing Irradiation

1. Free-radical polymerization

$RX + e^- \longrightarrow RX^*$ Excitation

$RX^* \longrightarrow R\bullet + X\bullet$ Free radicals

$M + R\bullet \longrightarrow RM\bullet + M \longrightarrow RMM\bullet \longrightarrow RM_x$

M – Monomer

R• – Free radical

2. Cross-linking

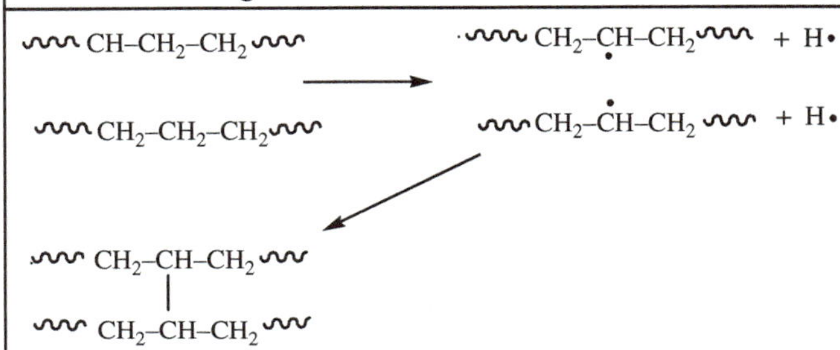

3. Main chain scission—polymer degradation

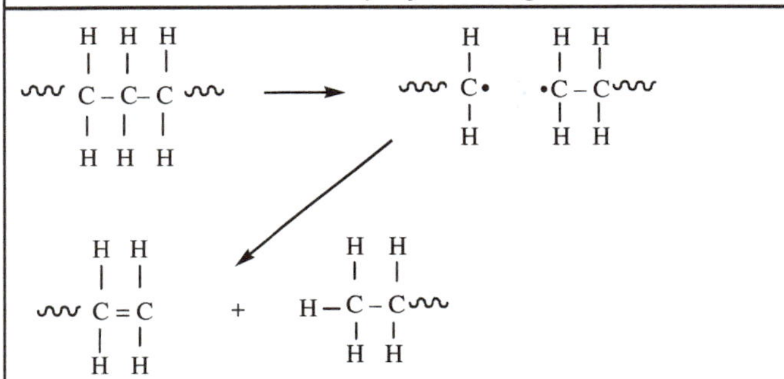

Disproportionation

4. Grafting

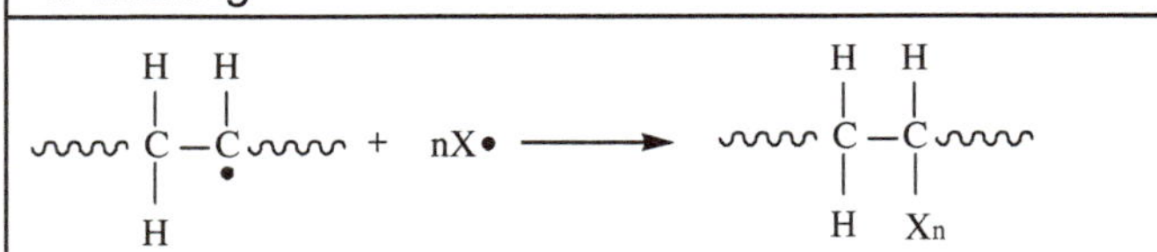

X-monomer

A similar method for polymers that cross-link is allowing calculation of $G(X)$ from the *gel point*, a point at which a polymer becomes insoluble. Knowing the average molecular weight of a polymer (Mw) and its "dose" at gel (d_g), one can determine G_{gel}, thus arriving at comparative $G(X)$ values

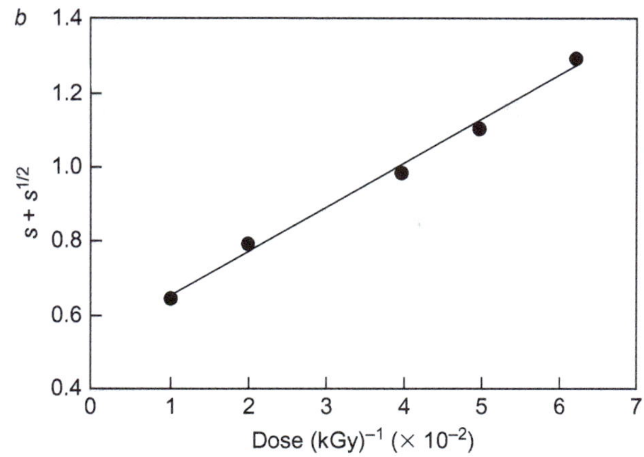

Figure 2.4 Charlesby–Pinner Plot.

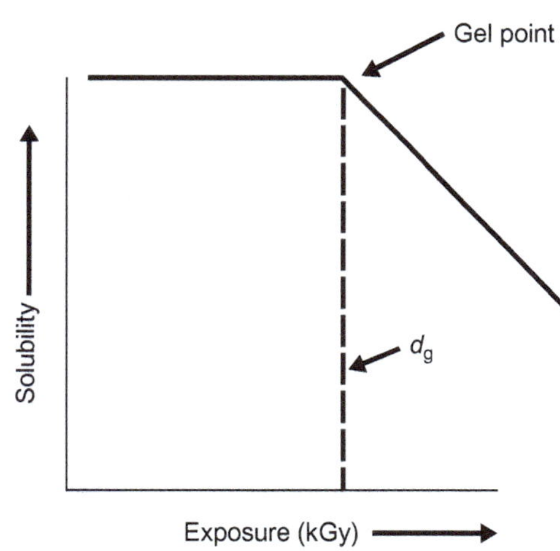

Figure 2.5 Illustration for the Determination of G_{gel}.

for various polymers. The dose at gel (d_g) can be determined by a series of solubility tests using incremental increases in exposure (dose), as illustrated in Figure 2.5 [10]. For some polymers, such as very high melt index polyethylenes (i.e., with low molecular weight), the competing cross-linking and scissioning events cancel each other out so that no effect is observed. For polymers with ring structures in which the absorbed radiation is believed to resonate within the carbon ring, the $G(X)$ and $G(S)$ values are very low, typically at the order of 10^{-2} [10].

The Charlesby–Pinner equation has been well used to evaluate both the *G*-values of many polymers. However, in some instances, its plot does not give a straight line. This is related to the assumptions used for the derivation of the Charlesby–Pinner equation. These assumptions are:

- The initial molecular weight distribution is random.
- The degree of cross-linking and main chain scission is proportional to the radiation dose.
- The structure of polymers does not affect the cross-linking and main chain scission.

For that reason, several modifications have been proposed that take the real conditions into consideration [11–14]. The details involved are beyond the scope of this publication. The reader will find a thorough discussion of this subject in Ref. [1], pp. 40–42.

Polymers can be classified into two groups according to their response to ionizing radiation. One group exhibits predominant cross-linking; the other, predominant chain scission (Table 2.3) [15]. *G*-values for selected polymers are listed in Table 2.4.

Evidently, the response to ionizing radiation depends on the structure of the polymer. Numerous studies have been performed to establish this correlation. One study [16] suggests that when the monomeric unit of the polymer contains at least one α-hydrogen, such as polyethylene and polyvinylidene fluoride, cross-linking will take place; if not, the main chain will degrade. Another study [17] proposed that vinyl polymers with two side chains attached to a single backbone carbon, i.e., ($-CH_2-CR_1R_2-$), such as polymethylmethacrylate, or perhalogenated polymers, such as polytetrafluoroethylene will degrade. Those polymers with an attached single or no side chain ($-CH_2-CR_1H-$ or $-CH_2-CH_2-$) will cross-link. Polymers with pending methyl group, such as polyisobutylene and polypropylene, are subject to degradation when exposed to ionizing radiation. Aromatic polymers with benzene rings in the main chain or pending benzene rings, such as polystyrene or polycarbonate, are usually resistant to ionizing radiation [1, p. 18].

The correlation between the heat of polymerization and the tendency to degrade or cross-link was discussed by Wall [18]. Currently, the generally accepted mechanism involves the cleavage of the C–H bond on one polymer chain to form a hydrogen atom, followed by abstraction of a second hydrogen atom from a neighboring chain. The two hydrogen atoms form the H_2 molecule. Then the two adjacent polymeric radicals combine to form a cross-link leading to branched chains until ultimately a three-dimensional polymer network is formed when each polymer chain is linked to another chain.

Table 2.3 Classification of Polymers According to Their Response to Ionizing Radiation

Polymers Predominantly Cross-Linking	**Polymers Predominantly Degrading**
Polyethylene	Poly(tetrafluoroethylene)
Polypropylene	Poly(α-methylstyrene)
Polystyrene	Poly(vinylidene chloride)
Poly(vinyl chloride)	Poly(vinyl fluoride)
Poly(vinyl alcohol)	Polychlorotrifluoroethylene
Poly(vinyl acetate)	Polytetrafluoroethylene
Poly(vinyl methyl ether)	Polyacrylonitrile
Polybutadiene	Polyvinylbutyral
Polychloroprene	Poly(methyl methacrylate)
Copolymer of styrene and acrylonitrile	Polymethacrylonitrile
Natural rubber	Polyoxymethylene
Chlorinated polyethylene	Poly(propylene sulfide)
Chlorosulfonated polyethylene	Poly(ethylene sulfide)
Polyamides	Cellulose
Polyesters	Polyalanine
Polyurethanes	Polylysine
Polysulfones	Polyisobutylene
Polyacrylates	DNA
Polyacrylamides	
Polydimethylsiloxane	
Polydimethylphenylsiloxane	
Phenol–formaldehyde resins	
Urea–formaldehyde resins	
Melamine–formaldehyde resins	

In contrast, scission is the opposite process of cross-linking in which the rupturing of the C–C bond occurs. Cross-linking increases the average molecular weight, whereas scission reduces it. If the energy of radiation is high, chain breakage occurs as a result of the cleavage of the C–C bond.

Table 2.4 *G*-Values for Selected Polymers

Polymer	*G*(*X*) Value	*G*(*S*) Value	Ratio *G*(*X*)/*G*(*S*)
Polybutadiene	5.3	0.53	10
Natural rubber	1.3–3.5	0.1–0.2	7.4
High-density polyethylene	0.96	0.19	5.05
Low-density polyethylene	1.42	0.1–0.2	2.95
Polyvinylidene fluoride	0.6–1.0	0.3–0.6	3.33
Polymethyl methacrylate	0.45–0.52	0.05	6.6
Polyamide 6/6	0.5–0.9	0.7–2.4	0.71
Atactic polypropylene	0.4–0.5	0.3–0.6	0.91
Isotactic polyethylene	0.16–0.26	0.29–0.31	0.67
Butyl rubber	<0.5	2.9–3.7	0.18
Polyisobutylene	0.05–0.5	5	0.05
Polytetrafluoroethylene	0.1–0.3	3.0–5.0	0.05

In an oxygen atmosphere, however, the mechanistic process of scission proceeds in an indirect manner. The polymeric-free radicals are generated by radiation and the addition of oxygen to the polymeric radicals forms peroxidic species, which on decomposition form shorter chains.

It can be concluded that it is very difficult to predict the result from polymer macrostructure, but it is relatively easy to measure the secondary species generated on irradiation by using known analytical techniques, such as measuring swelling, tensile tests, and NMR analysis.

Cross-linking and chain scission are two competing processes that always coexist under irradiation. The overall effect depends on which of the two is predominant at a certain time [1, p. 18]. Whenever $G(X)$ is greater than $G(S)$, the overall result is cross-linking; in the opposite case, the overall result is degradation. The value of $G(X)$ or $G(X)/G(S)$ is used to evaluate the efficiency of radiation cross-linking. Three-dimensional networks are formed when $4G(X)$ is larger than $G(S)$. The *G*-values for a polymer change with radiation conditions, such as absorbed dose and the temperature.

Studies performed on elastomers have shown that the cross-linking efficiency is relatively low. $G(X)$ values of 1 and approximately 3 were found for natural rubber and polybutadiene respectively when irradiated at room temperature. Temperature was shown to have positive effect in some cases [19, 20].

Gas evolution happens in almost all polymers under irradiation. The gas formed is mainly hydrogen, and also methane, carbon monoxide, carbon dioxide, and others [21]. Gas yields from selected polymers are given in Table 2.5.

Table 2.5 Gas Yields from Selected Polymers

Polymer	Gas Yield in ml/g (STP) at 10^4 kGy
Polyethylene	70
Polystyrene	1.5
Poly(α-methylstyrene)	1.5
Natural rubber	7
Styrene–butadiene rubber (SBR)	7
Acrylonitrile–butadiene rubber (NBR)	5
Isobutylene–isoprene rubber (IIR, butyl rubber)	17
Silicone rubber	20
Polyamide	25
Phenol–formaldehyde resin (no filler)	3
Phenol–formaldehyde resin (with cellulose filler)	17
Phenol–formaldehyde resin (with mineral filler)	<2
Aniline–formaldehyde resin	~2
Melamine–formaldehyde resin (with cellulose filler)	10
Urea–formaldehyde resin (with cellulose filler)	17
Poly(methylmethacrylate)	35
Poly(ethylene terephthalate)	3
Cellulose acetate resin	17
Cellulose nitrate resin	105

Radiation-induced electrical conductivity is found in most plastics due to the formation of ions, and the induced current is a function of the dose rate. The conductivity usually decays exponentially over periods of days or months [21].

If the polymeric material is in the glassy state, i.e., below the glass transition temperature (T_g), or has some crystallinity, trapped electrons and trapped radicals can be produced upon irradiation with either EB or γ-ray irradiation. Phenomena such as thermoluminescence, electrical conductivity, color changes in the polymer, and imperfections in the crystals have been attributed to ionic species. Trapped electrons have been identified in γ-irradiated polyethylene [22].

In general, amorphous materials do not have the tendency to produce trapped electrons. Polymeric materials that contain trapped electrons undergo photobleaching when exposed to near-IR light ($\lambda > 1000$ nm) [23].

Any effect of irradiation, in general, is increased with increasing temperature. At temperatures below T_g, a significant number of stable radicals are formed (radical trapped in the glassy state) and, in general, cross-linking is reduced due to the immobility in the glassy state. At temperatures above T_g, the tendency to cross-link is usually increased, although chain scission processes will also increase.

Semicrystalline polymers contain both crystalline and amorphous states. In general, the major effect of irradiation, either EB or γ-rays, on the crystalline region is to cause some imperfections. At high levels of irradiation, the original crystalline structure tends to be progressively destroyed and this process is nearly always accompanied by a drop in the crystalline melting point T_m. (An example is poly(ethylene terephthalate) which shows a decrease in melting point of approximately 25°C after irradiation with an absorbed dose of 20,000 kGy) [24].

References

[1] Makuuchi K, Cheng S. Radiation processing of polymer materials and its industrial applications. Hoboken, NJ: John Wiley & Sons; 2012.

[2] Bradley R. Radiation technology handbook. New York, NY: Marcel Dekker; 1984. p. 17.

[3] Samuel AH, Magee JL. J Chem Phys 1953;21:1080.

[4] Clegg DW, Collyer AA, editors. Irradiation effects on polymers. London: Elsevier; 1991.

[5] Singh A, Silverman J, editors. Radiation processing of polymers. Munich: Carl Hanser Verlag; 1992.

[6] Mehnert R. Radiation chemistry: radiation induced polymerization in Ullmann's encyclopedia of industrial chemistry, vol. A22. Weinheim: VCH; 1993. p. 471.

[7] Garratt PG. Strahlenhärtung. Hannover: Curt R.Vincentz Verlag; 1996. p. 61, In German.
[8] L'Annunziata M, Baradei M. Handbook of radioactivity analysis. Waltham, MA: Academic Press; 2003. p. 58.
[9] Charlesby A, Pinner SH. Proc R Soc London Series 1959;A249:367.
[10] Industrial Radiation Processing with Electron Beams and X-rays. International Atomic Energy Agency. Vienna, Austria; 2011. p. 33. <www.iaea.org>.
[11] Olejniczak J, Rosiak J, Charlesby A. Radiat Phys Chem 1991;37:499.
[12] Rosiak J. Radiat Phys Chem 1998;5:13.
[13] Sun J. Radiat Phys Chem 2001;60:445.
[14] Zhang L, Zhou M, Chen D. Radiat Phys Chem 1994;44:303.
[15] Clough R. In: Kroschwitz JI, editor. Encyclopedia of polymer science and engineering, vol. 13. New York, NY: John Wiley & Sons; 1988.
[16] Miller AA, Lawton EJ, Balwit JS. J Polym Sci 1954;14:503.
[17] Charlesby A. Atomic radiation and polymers. New York, NY: Pergamon Press; 1960. p. 184.
[18] Wall LA. J Polym Sci 1955;17:141.
[19] Pearson RW, Bennett JV, Mills IG. Chem Ind (London) 1960:1572.
[20] Kozlov VT, Yevseyev AG, Zubov PI. Vysokomol Soed 1969;A11 (10):2230.
[21] Willis PB. Survey of radiation effects on materials, a presentation at the OPFM Instrument Workshop. Monrovia, CA; 2008 [OPFM5 Outer Planet Flagship Mission].
[22] Keyser RM, Tsuji K, Williams F. Macromolecules 1968;1:289.
[23] Dawes K, Glover LC. In: Mark JE, editor. Physical properties of polymers handbook. Woodbury, NY: American Institute of Physics; 1996. p. 562.
[24] Kusy RP, Turner DT. Macromolecules 1978;4:337.

Recommended Further Reading

Makuuchi K, Cheng S. Radiation processing of polymer materials and its industrial applications. Hoboken, NJ: John Wiley & Sons; 2012.

Industrial Radiation Processing with Electron Beams and X-rays, International Atomic Energy Agency, Vienna, Austria; May 2011.

Gamma Irradiators for Radiation Processing, International Atomic Energy Agency, Vienna, Austria.

Drobny JG. Radiation technology for polymers. 2nd ed. Boca Raton, FL: CRC Press; 2010.

L'Annunziata M, Baradei M. Handbook of Radioactivity analysis. Academic Press; 2003.
Singh A, Silverman J, editors. Radiation processing of polymers. Munich: Hanser Publishers; 1992.
Taniguchi N, Ikeda M, Miyamoto I, Miyazaki T. Energy-beam processing of materials. Oxford: Clarendon Press; 1989.
Cuomo JJ, Rossnagel SM, Kaufman HR. Handbook of ion beam processing technology. Park Ridge, NJ: Noyes Publishing; 1989.
Feynman R, Leighton R, Sands M. Feynman lectures on physics, vol. 1. Addison-Wesley; 1963.
Charlesby A. Atomic radiation and polymers. Oxford, UK: Pergamon Press; 1960.

3 Equipment Used for Radiation Processing of Polymers

OUTLINE

Currently, there are three types of equipment generating ionizing radiation that are used for processing polymers:

1. γ-Ray sources
2. EB accelerators
3. X-ray reactors

Each of them is suitable for specific types of applications, because there are differences in their properties, such as depth of penetration, dose rate, dose distribution, temperature effects, throughput, degree and type of cross-linking process, types of cross-links, extent of chain scission, and degradation. Comparison of γ-ray and EB sources is discussed in Section 1.2.1.

3.1 γ-Ray Irradiation Equipment

Today, cobalt-60 is almost solely used as the γ-radiation source for industrial applications mainly because of its easy production method and its insolubility in water. In the irradiation process, γ-radiation offers the following advantages:

- Treated product can be used immediately
- Minimal rise in product temperature during the process

Drobny: Ionizing Radiation and Polymers.
DOI: http://dx.doi.org/10.1016/B978-1-4557-7881-2.00003-1

- High penetration
- Very precise and reproducible treatment process
- Easy control of the process (only dose to be controlled).

The *radionuclide* cobalt-60 ($^{60}Co_{27}$) is the most commonly used source of γ-radiation for radiation technology, both for industrial and medical purposes. It is prepared from the nonradioactive cobalt-59 ($^{59}Co_{27}$) using neutron irradiation in a specifically designed nuclear power reactor. The process takes 18–24 months, depending on the neutron flux at the location.

Cobalt-60 is used mainly in the form of rods or pencils (Figure 3.1). The pencils are double-skinned capsules about 45 cm (18 in) long, with a diameter of around 8 mm (0.3 in) made of stainless steel. These capsules encase the radioactive material. The pencils are grouped in subassemblies which are then placed into predetermined positions in source modules. These source modules are distributed over the source rack of the industrial irradiator (Figure 3.2). The modules are stored in a deep water storage pool when not in use.

When the modules are in use, cobalt decays in the following fashion: First $^{60}Co_{27}$ decays into excited $^{60}Ni_{28}$ by beta decay with the emission of an electron with the energy of 0.31 MeV. Then the excited $^{60}Ni^{*}_{28}$ drops into the ground state, $^{60}Ni_{28}$, by emitting two γ-rays (photons) in succession (1.17 MeV and then 1.33 MeV). This path is followed 99.8% of the time (Figure 3.3). The simplified equation for the process is

$$^{60}Co_{27} \rightarrow \underset{\text{Excited}}{^{60}Ni^{*}_{28}} \rightarrow \underset{\text{Ground state}}{^{60}Ni_{28}}$$

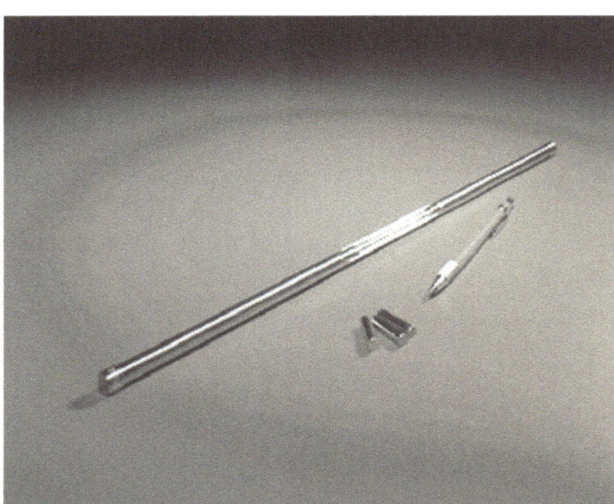

Figure 3.1 Cobalt-60 Source ("Pencil"). *(Courtesy of Nordion (Canada) Inc.)*

Figure 3.2 Source Rack Module. *(Courtesy of Nordion (Canada) Inc.)*

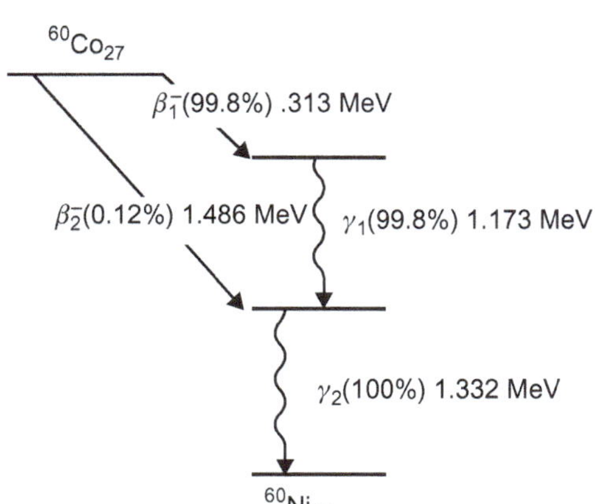

Figure 3.3 Decay Scheme of Radionuclide Cobalt-60.

The two γ-ray photons from the transition of cobalt-60 to nickel-60 (1.17 and 1.33 MeV) are mainly responsible for the radiation processing with the cobalt-60 γ-irradiators.

In radiation processing, a product or material is intentionally irradiated to preserve, modify, or improve its characteristics. This process is carried out by placing the product in the vicinity of a radiation source for a fixed time interval whereby the product is exposed to radiation emanating from the source. A fraction of the radiation energy that reaches the product is absorbed by the product, the amount depending on its mass and composition, and the time of exposure. For each type of product, a certain amount of radiation energy is needed to realize the desired effect in the product; the exact value is determined through research.

In line with decay, the strength (the radioactivity level) of the cobalt source decreases, such that the decrease amounts to 50% in about 5.27 years or approximately 12% in 1 year. Additional pencils of cobalt-60 are added periodically to the source rack to maintain the required capacity. Cobalt-60 pencils are eventually removed from the irradiator at the end of their useful life, which is typically 20 years. Generally, they are returned to the supplier for reuse or disposal. In about 50 years, 99.9% of cobalt-60 would decay into nonreactive nickel. For the currently available commercial γ-irradiators, typically 30% of the energy emitted by the radiation source is usefully absorbed by the product [1].

A commercial γ-ray irradiation facility (Figure 3.4) includes the irradiation room, shielded storage room for the radiation source rack, irradiation shield surrounding the irradiation room, control console, product containers, product transport system, control and safety interlock system, and areas for loading and unloading of products. The irradiation shield generally consists of a concrete wall thick enough (normally 2 m in thickness) to attenuate the radiation emanating from the source. The concrete wall is constructed as a maze so as to permit movement of the product, yet significantly reduce the scattered radiation to reach the control console.

There are several types of irradiators available commercially. The size of an irradiator depends on the intended use. The design principles for all irradiators are to maximize radiation energy utilization, to provide relatively uniform dose in the product, and to ensure safe and easy operation. Essentially, there are two major types of irradiators:

1. Self-contained irradiators
2. Panoramic irradiators.

Self-contained irradiators are specially designed for research and applications needing small doses and relatively small throughputs, such as blood irradiation and reproductive sterilization of insects for pest control. The

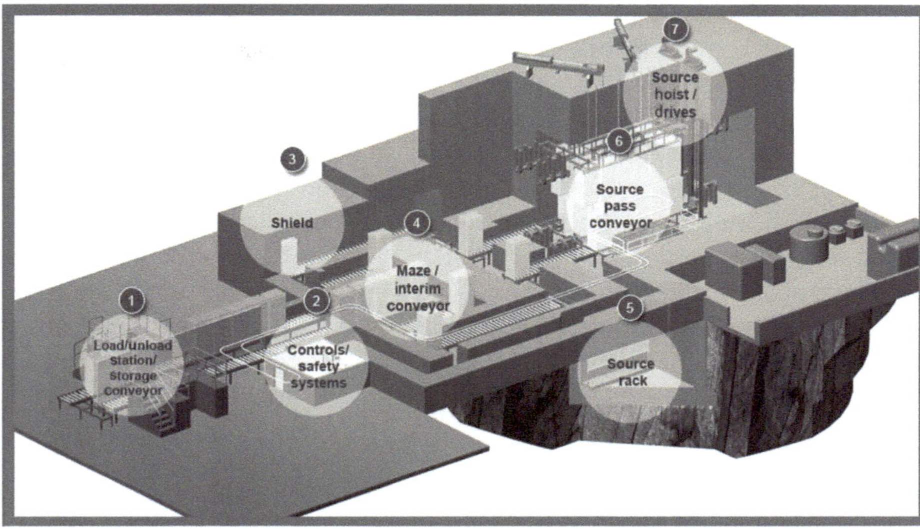

Figure 3.4 Commercial γ-Ray Irradiation Facility. *(Courtesy of Nordion (Canada) Inc.)*

principal advantage of such radiators is that they are easily installed and operated and provide high dose rate and good dose uniformity. These irradiators are classified by International Atomic Energy Agency (IAEA) as Category I (dry storage) and Category III (wet storage) [2].

Panoramic irradiators are used as pilot scale and commercial scale irradiations. The source here consists of several cobalt-60 pencils arranged in either a plane or a cylinder. These irradiators are classified by IAEA as Category II (dry storage) and IV (wet storage) [3].

The dose rate in the product is directly proportional to the installed activity of the source. The operator controls the absorbed dose delivered to the product by adjusting the time it is exposed to radiation. For the currently available commercial γ-irradiators, typically 30% of the energy emitted by the radiation source is usefully absorbed by the product. There are several designs, depending on the method of irradiation and the product involved (batch irradiators, product overlap, source overlap irradiators, or pallet irradiators).

Process dose, that is the dose needed to achieve a desired effect in the product, is determined through radiation research, which involves determining the dose–effect relationship for the product/effect [1]. Generally, the outcome of such research is identification of two dose limits: the lower dose limit sets the minimum dose that is required to achieve the desired effect on the product and the upper dose is set to assure that the radiation will not adversely affect the functional quality of the product. The ratio of the upper dose limit to the lower dose limit may be referred to as *dose limit ratio* [1; p. 16].

As the radiation proceeds through a product, its intensity gradually decreases (see Section 1.2.1). The decrease of the intensity is referred to as *depth-dose distribution*. The rate of decrease depends on the composition and density of the product, as well as the energy of the γ-radiation. Besides the variation in depth, there is also dose variation in the lateral direction. This variation depends on the geometry of the irradiation. Both types of dose variation contribute to the nonuniformity of the dose delivered to the product. One accepted method of describing this nonuniformity of dose is the concept of *dose uniformity ratio* (DUR), which is the ratio of the maximum dose in a product container to the minimum dose in the container. This ratio should be as close to unity as possible.

3.2 EB Irradiation Equipment

Commercial EB (electron beam) processors available today vary in accelerating voltage range from about 100 kV up to about 10 MV. The original development of EB technology in the 1950s concentrated on machines with accelerating voltages in the range from 1 to 2 MV with beam powers of 5–10 kW suitable primarily for the cross-linking of plastic materials. During the 1960s, the successful applications required even higher voltages up to 3 and 4 MV with power output levels up to 100 kW [4]. Lower voltage machines in the range below 300 kV were developed in the 1970s. They were designed without a beam scanner, which was used in the higher voltage equipment. The EB in the lower voltage range was delivered as a continuous "shower" or "curtain" across the entire width of the product. The main applications of the lower voltage machines were coatings and other thin film layers [4]. The most recent designs use voltages in the range 70–125 kV with electron energies suitable for the processing of thin layers of inks and lacquers less than 25 μm thick.

3.2.1 Design and Operation of Industrial EB Reactors

The principle of producing high-energy electrons is very simple. The electrons are emitted in a vacuum by a heated cathode and accelerated in the electrostatic field applied between cathode and anode. Acceleration takes place from the cathode, which is connected to a negative high-voltage potential, to the grounded accelerator window as anode. Usually an electron optical system is used to focus the accelerated electrons to the accelerator window plane.

The energy gain of the electrons is proportional to the accelerating voltage. It is expressed in electron volts (eV), i.e., the energy which a particle of unit charge gains by passing a potential difference of 1 V. The electrons leave the vacuum chamber and reach the process zone if their energy is high enough to penetrate the 6–14 μm thick titanium window foils used in the lower energy range and 40–50 μm thick foils used in the higher energy range.

As pointed out in Section 2.2, stopping the high-energy electrons by a material generates X-rays. For that reason, the electron accelerator and process zones have to be shielded to protect the operator. For electrons with energies up to 300 keV, self-shielding with lead cladding up to about 1 in thick is sufficient. For systems where electrons with higher energy are generated, the equipment is shielded by a concrete or steel vault built around it as shown in Figure 3.5.

The basic electrical parameters of an EB processor are its *acceleration voltage*, the *EB current*, and the *EB power*. The ratio of EB power and of the input electrical power defines the efficiency of an electron accelerator [5]. The acceleration voltage determines the energy of the electrons as pointed out in Section 2.2.

3.2.1.1 Particle Accelerators

Particle accelerators produce electrically charged particles with speeds slightly over 6000 miles/s to speeds approaching the speed of light. Accelerated particles are generally formed into a beam and directed at a material called a target [6]. As this beam passes through the window material, some of the particles are deflected and may interact with atomic nuclei. The final effect depends on the energy and nature of the incident particle. A particle accelerator is depicted schematically in Figure 3.6 [6].

There are several designs of accelerators for industrial use. They are mainly used to accelerate electrons, although they can be adapted to accelerate ions as well. For the purpose of this publication, accelerators of electrons will be discussed at some length.

Essentially, there are two types of electron accelerators used in industrial EB applications. *Direct accelerators* are also referred to as *potential drop accelerators* and require the generation of high electrical potentials equal to the final energies of the accelerated electrons. *Indirect accelerators* produce high electron energies by repetitive application of time-varying electromagnetic fields (J. Chrusciel, private communication). Direct accelerators are widely used for low- and medium-energy applications since they are able to

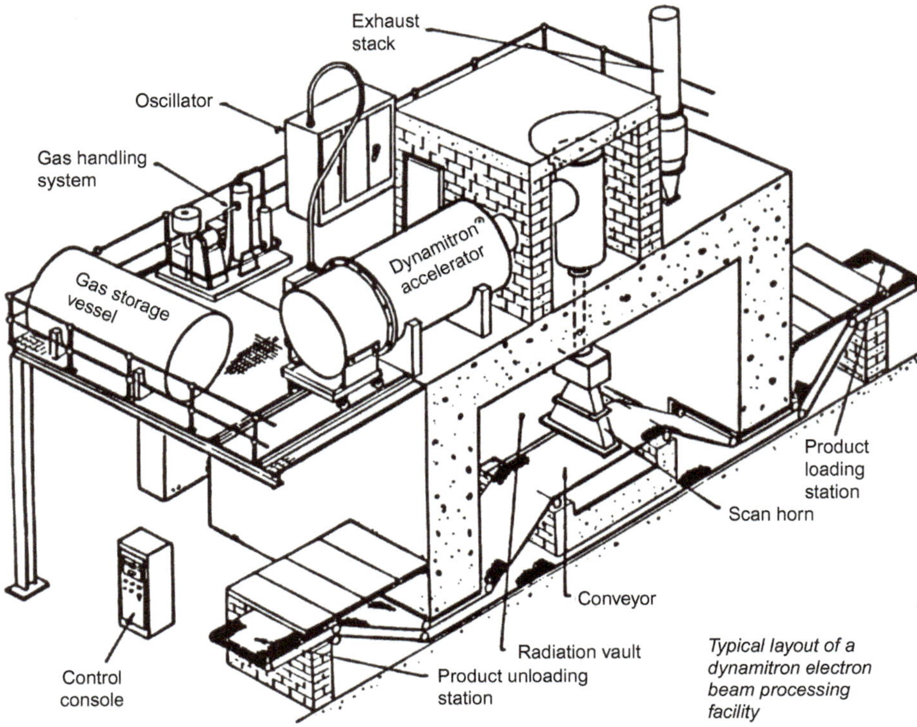

Figure 3.5 Traditional Design of a High-Energy EB Unit. *(By permission of Elsevier Science.)*

provide continuous EBs with high average current and power ratings, which translate to high processing rates. For higher energies (about 5 MeV), indirect accelerators using microwave radiation, very-high-frequency (VHF) radiation, or pulsed electrical power are considered to be more suitable. Modern accelerators can generate beam powers up to several hundred kilowatts and electron energies up to 15 MeV.

3.2.1.1.1 Direct Accelerators

In principle, a direct electron accelerator consists of a high-voltage generator connected to an evacuated acceleration system. The different direct accelerators currently used employ similar methods for electron emission, acceleration, and dispersion; the differences are in the design of their voltage generators.

The source of electrons is a thermionic cathode, which is almost always a tungsten or thoriated tungsten wire, although tantalum wire or lithium hexaboride pellets are also used as cathodes. Electron emission is most frequently controlled by the variation of temperature of the cathode, but this can be done

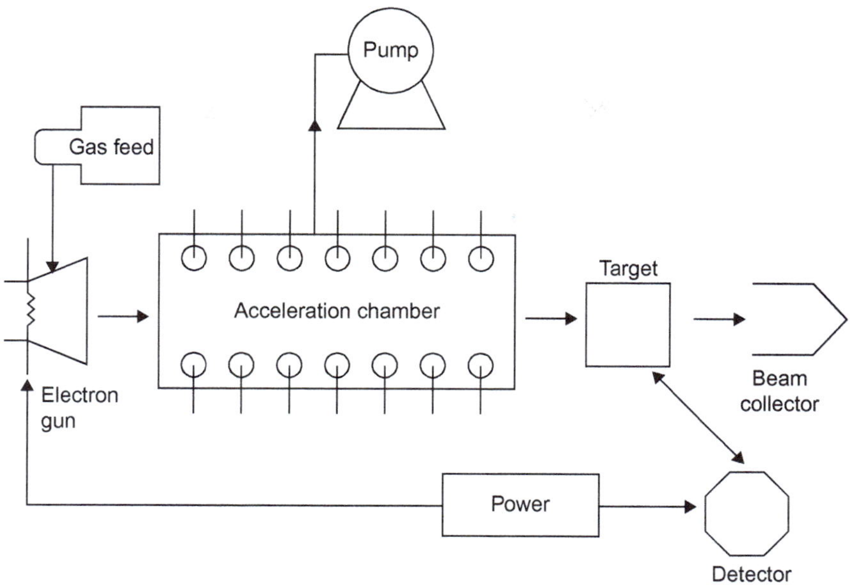

Figure 3.6 Schematic of a Particle Accelerator.

also by a grid with variable voltage. The beam of electrons is extracted, focused, and accelerated by an internal electric field, determined by a series of electrodes or dynodes with intermediate potentials obtained from a resistive voltage divider. The electrons gain energy continuously throughout the length of the acceleration tube. After acceleration, the concentrated EB is dispersed by scanning with a time-varying magnetic field with a sweep rate of at least 100 Hz. The divergent beam so formed expands within an evacuated scan horn and then emerges into the air through a thin metallic foil. This foil is usually made from titanium, but other metals, such as aluminum alloys or titanium alloys, are used. In order to minimize the loss of the electron energy, the thickness of a titanium foil is kept in the range between 25 and 50 μm for electron energies above 0.5 MeV and between 6 and 15 μm for energies below 0.3 MeV. The windows of the scanning horn may be as wide as 2 m (80 in).

Electrostatic Generators Electrostatic generators are based on the principle that high electric potentials can be produced by mechanically moving static charges between low-voltage and high-voltage terminals. The various designs have used endless rubber belts, chains or metallic electrodes with insulating links, and rotating glass cylinders. Compressed gases, e.g., nitrogen, sulfur hexafluoride (SF_6), Freon, and carbon dioxide, are used for high-voltage insulation to reduce the size of the equipment. One example of such a device is a Van de Graaf generator (Figure 3.7), a belt-driven machine originally developed in the 1930s for research in nuclear physics; industrial versions capable of delivering several kW of EB power with energies up to 4 MeV were developed in the 1950s. It was used for EB cross-linking of plastic films, tubing, insulated wire, and biomedical applications [7]. Today, these machines are no longer competitive with modern high-power machines and are therefore seldom used in industrial applications.

Resonant Transformers Pulsed EBs produced by these devices are generally limited to an average current of about 5–6 mA since they operate at a relatively low frequency of 180 Hz. The voltage of such an accelerator is not rectified as it is in other industrial accelerators, and the high voltages (typically 1.0 and 2.0 MV) of commercial resonant transformers produce pulsating current at their resonant frequency. This pulsating current makes it difficult to achieve uniform dose in the irradiated material. These machines consist of a pressure tank, in which the iron-free resonant transformer and the discharge tube are placed. Secondary windings connected in series feed the high-voltage terminal. The system provides a beam only during the negative half cycle, with a voltage variation between zero and the peak value the machine

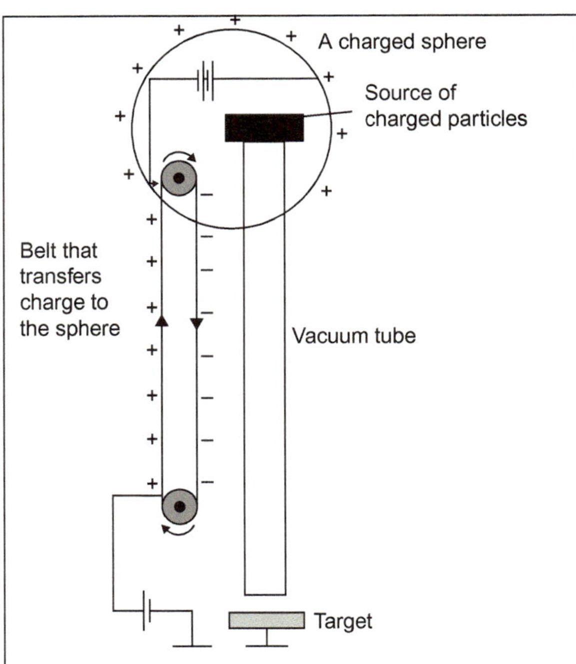

Figure 3.7 Schematic of the Van de Graaf Generator/Accelerator.

was designed to deliver. Sulfur hexafluoride gas is used for electrical insulation [8].

Iron Core Transformers Low-frequency transformers with iron cores connected to rectifier circuits can generate potentials up to approximately 1 MV. Traditionally, iron core transformers are insulated by oil, but the recent models use sulfur hexafluoride. The energy rating of these generators is in the range from 0.3 to 1 MeV with beam powers up to 100 kW [9,10].

Insulating Core Transformers Insulating core transformers (ICTs) were developed in the late 1950s and early 1960s to replace Van de Graaf electrostatic generators to provide more irradiation power for higher processing rates. An ICT consists of a three-phase transformer with multiple secondary windings that are energized serially by iron core segments. These segments are separated by thin sheets of dielectric material. Low-voltage, low-frequency AC power is converted into high-voltage DC power by using rectifier circuits connected to the nearest core segment. This way the electrical stresses in the system are minimized. ICT designs are capable of higher voltage ratings than conventional iron core transformers [9]. The insulating medium is sulfur hexafluoride gas and the line to beam power efficiency is close to 85% [11–14]. The terminal voltage of the ICT power supply is connected to an accelerator tube. The EB for industrial applications starts from a tungsten filament and is accelerated within the evacuated tube as in other DC accelerators.

ICT machines produced recently have energy ratings from 0.3 to 3.0 MeV and beam power capabilities up to 100 kW. Nearly 180 of these machines, the majority of them rated for less than 1 MeV, have been installed as of the early 1990s. They are used mainly for cross-linking of heat-shrinkable film, plastic tubing, and electric wire [9].

Cockcroft–Walton Generators These are essentially cascade generators. In this type of electron accelerator, high voltage is produced by the incremental movement of electric charge. The high-voltage system is a condenser-coupled, cascade rectifier, which converts low-voltage, medium-frequency (3 kHz) AC power into high-voltage DC power. The capacitative coupling circuits are connected in series. Sulfur hexafluoride is used as insulation [15]. Cockroft–Walton generators are made in different designs. Three-phase rectifier circuits are used for low-energy, high-current applications, conventional single-phase cascades for energies between 1 and 3 MeV, and balanced two-phase systems for energies between 1 and 5 MeV with beam power ratings up to 100 kW. Electrical efficiencies are about 75% [16].

Dynamitrons In the Dynamitron type of accelerator (Figure 3.8), the high-voltage DC power is generated by means of a cascade rectifier circuit

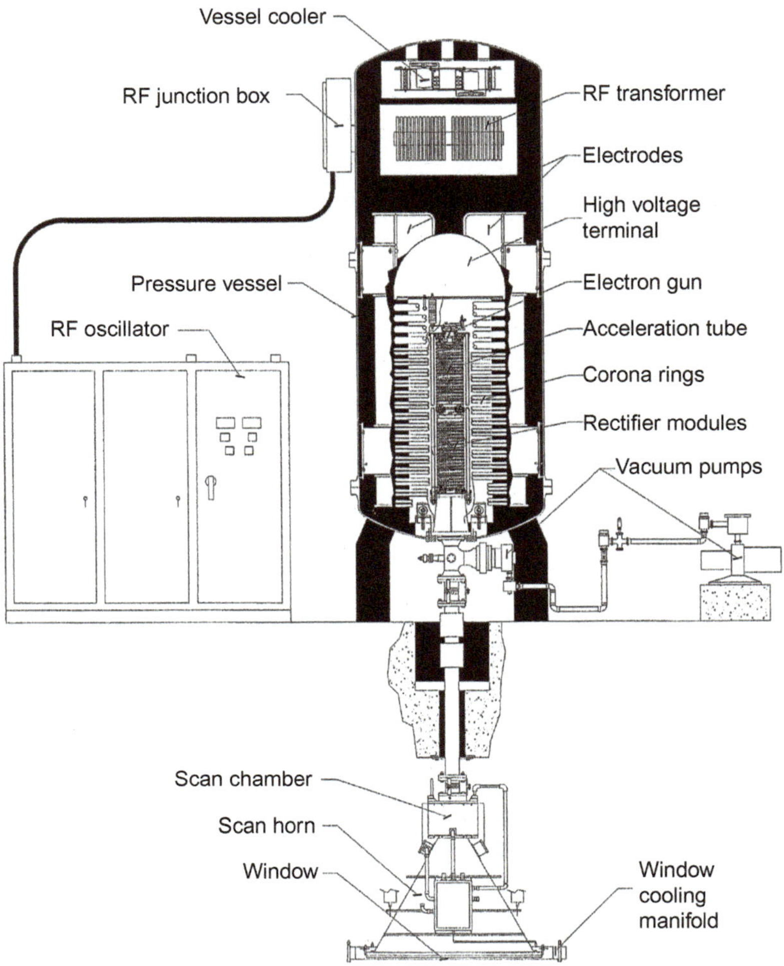

Figure 3.8 Schematic of a Direct Accelerator (Dynamitron). *(Courtesy of IBA Industrial.)*

energized by high-frequency (100 kHz) AC power. The rectifiers are driven in parallel by a pair of large semicylindrical electrodes, which surround the high-voltage column. This arrangement enhances the reliability at the maximum voltage by eliminating the large, high-voltage condensers used in a series-coupled Cockcroft–Walton generator. Because of power loss in the high-frequency oscillator, employing a triode vacuum tube, the electrical efficiency of a Dynamitron accelerator is about 60% [17,18]. These machines have been made with energy ratings from 0.5 to 5 MeV and beam power ratings up to 250 kW. More than 150 Dynamitrons, the majority of which have energy ratings above 1 MeV, are used for the cross-linking of polymeric materials [9]. Some of the larger machines with energy ratings above 3 MeV are also used for sterilization of medical products by energetic electrons [19]. The largest model, which can deliver 250 kW of beam power at energies in the range from 3 to 5 MeV, is equipped with a high-power target for X-ray processing [20,21].

3.2.1.1.2 Indirect Accelerators

This type of accelerator produces high-energy electrons by injecting short pulses of low-energy electrons into a copper waveguide, which contains intense microwave radiation. When the injection phase is at its optimum, the electrons are able to gain energy from the alternating electromagnetic field. Their final energy depends on the average strength of the field and the length of the waveguide [16]. Such an accelerator is referred to as a *microwave linear accelerator* or *linac* and is the most prevalent type of indirect accelerator [16]. It is shown schematically in Figure 3.9.

Linacs operate with much lower average beam current and power levels than most DC accelerators to attain sufficiently high electron energy. The relatively low voltage of the microwave component and the use of a grounded copper waveguide result in smaller size when compared to DC machines with the same electron energy. Their disadvantage is a low overall electrical efficiency ranging from 20% to 30% [22].

Low-power linacs are mostly used in cancer therapy and industrial radiography, whereas medium-power linacs are utilized in radiation processing.

Traveling Wave Linacs In traveling wave (TW) linacs, the microwave power is injected in one end of the waveguide and propagates to the other end. At that point, any residual microwave power is dissipated in a resistive

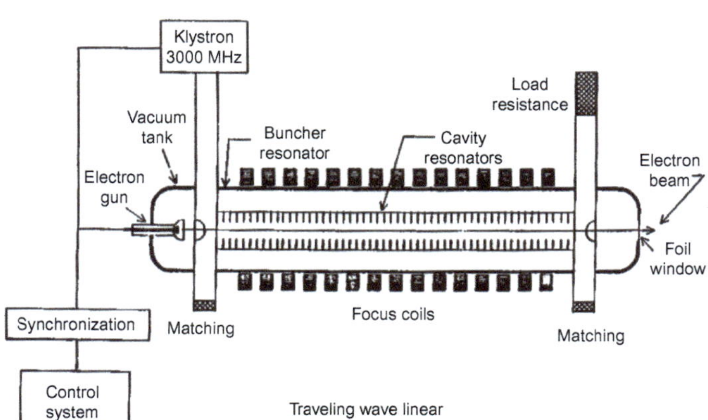

Figure 3.9 Schematic of an Indirect Accelerator (Microwave Linac).

load. Low-energy electrons are also injected with the microwave power and travel along in step with the moving electric wave gaining energy from it continuously.

The peak microwave power must be several MW in order to obtain energy gains of several MeV per meter and the system must be operated in short, repetitive pulses to keep the average microwave power down to reasonable levels. The EB current during the pulse ranges from 0.1 to 1.0 A for an average beam power of 10–20 kW at the electron energy of 10 MeV [23]. Although with optimal beam current, TW linacs can transfer microwave power with an efficiency ranging from 80% to 90% [24], because of the low efficiency of the microwave generation in a klystron, the overall line to beam power efficiency is usually below 30%.

Standing Wave Linacs A standing wave (SW) microwave linear accelerator consists of a linear array of resonant cavities that are energized by a common source of microwave power. These cavities are nearly isolated by webs with small-diameter apertures and the high-energy EB passes through these apertures. However, they are coupled through intermediate cavities, which stabilize the microwave phase relationship between the accelerating cavities.

SW linacs have higher electrical impedance or quality factor (Q) than TW linacs with similar energy and beam power ratings [16]. This can provide higher energy gain for the same length of waveguide or a shorter waveguide for the same energy gain, which is an advantage if the accelerator for the given application needs to be compact.

Resonant Cavity Accelerators Resonant cavity accelerators consist of several resonant cavities in series energized by a single S-band klystron using a microwave power distribution system [25]. Another system consists of a single VHF cavity energized by a triode tube that is less expensive than klystron. The resonant frequency of the latter is about 110 MHz, which is well below the microwave range [26]. Resonant cavity accelerators with electron energies of a few MeV are useful for irradiating thin polymeric products such as heat-shrinkable tubing and electrical wire [27].

Linear Induction Accelerators A linear induction accelerator (LIA) accelerates electrons by a series of single-turn, torroidal pulse transformers through which the beam passes. The energy gain in each stage is equal to the voltage applied to the primary winding. The beam acts as a secondary winding. The electrical impedance of an LIA is very low. This makes it suitable for accelerating high peak beam currents. In principle, the electrical efficiency of this accelerator can be substantially greater than that of a microwave linac [28].

Detailed information on designs, principles of operation, and performance characteristics of electron accelerators is available in several books [6,11,12,29,30].

Rhodotron Accelerators A Rhodotron is an electron accelerator based on the principle of recirculation of a beam in successive passes through a single coaxial cavity resonating in the VHF frequency range. This large-diameter cavity operates with a relatively low microwave field, which makes it possible to achieve continuous wave (CW) acceleration of EBs to high energies.

Rhodotron cavities are shaped as coaxial lines shorted on both ends and resonating in the half-wavelength mode 107.5 or 215 MHz. The beam crosses the diameter of the cavity in the median plane through successive passes (Figure 3.10) [31]. External dipole magnets are used to bend back the electrons emerging from the cavity and redirect them toward the cavity center. A high-power RF system using a tetrode tube produces the electric field allowing an energy gain of 0.83–1.17 MeV per crossing. Ten or twelve crossings of the cavity (which means nine or eleven bending magnets) are therefore required to obtain 10 MeV EBs at the exit of this accelerator. A very-high-power version can produce a 100 mA beam at 5–7 MeV with six passes. With this version, the average EB power is in the range of 500–700 kW, which can be used for producing intense X-ray (bremsstrahlung) beams. Since the electrons travel along a rose-shaped path, the name Rhodotron was chosen, derived from the Greek word "rhodos," which means "rose" [31].

The electron gun is located at the outer wall of the accelerating cavity and the electrons are injected into the cavity at the voltage of about 35–40 kV. The cavity is cooled by a water jacket on the inner coaxial conductor and at the end flanges and by discrete water channels along the outer

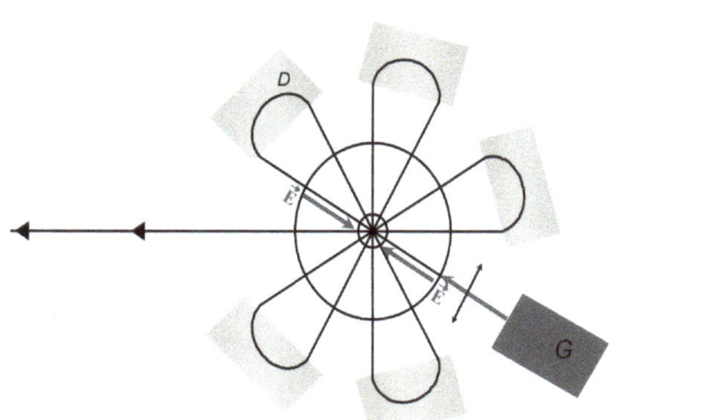

Figure 3.10 Schematic of a Rhodotron Accelerator; Principle of Operation. *(Courtesy of IBA Industrial.)*

diameter. The system is designed to operate with a 2 MW cooling tower up to an outside temperature of 35°C (95°F). Therefore, no water chiller is required [31]. The RF amplifier detects and follows changes in the resonant frequency of the cavity so that accurate control of the cavity temperature is not required.

The EB is of very high quality with low-energy spread and low angular divergence [32]. This simplifies the design and operation of external beam transport and scanning systems [33]. The beam delivery system includes a scanning horn with a vacuum-to-atmosphere window using a thin metal foil. An example of a Rhodotron with a scanning horn is shown in Figure 3.11.

The control system is based on an industrial programmable logic controller (PLC). It includes all software required for the completely automatic operation, maintenance, and troubleshooting of the accelerator [31].

In addition to providing EBs with 5 and 10 MeV electron energy outputs for usual EB applications, the Rhodotron can be used to produce *bremsstrahlung X-rays* from a metallic target that can be used for industrial applications [31,34–36]. The Rhodotron is particularly suitable for applications which need a powerful beam (30–200 kW) in the energy range of 1–10 MeV [37].

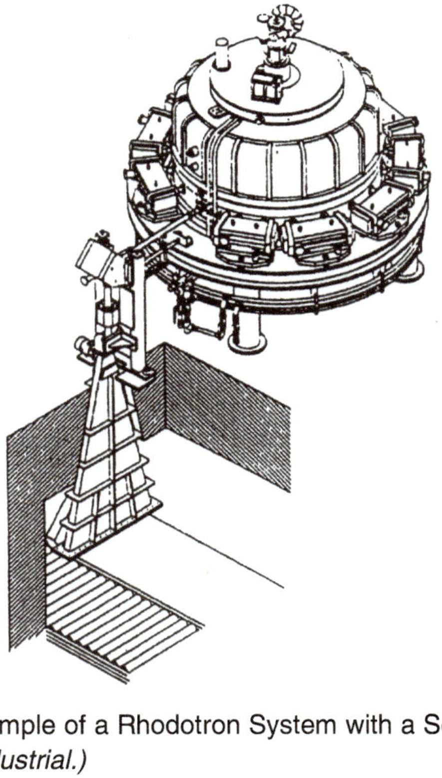

Figure 3.11 An Example of a Rhodotron System with a Scanning Horn. *(Courtesy of IBA Industrial.)*

ILU Accelerators ILU accelerators are developed and produced by the Budker Institute for Nuclear Physics (BINP) in Novosibirsk, Russia. They are essentially pulse linear high-frequency machines used for a wide range of industrial applications, including wire and cable, shrink tubes, cross-linking of pipes, and sterilization of medical goods and food, and in semiconductor manufacture [38]. Currently three models, ILU-6, ILU-8, and ILU-10, are available commercially and ILU-14 is reported to be in development [39]. The compact model ILU-8 has local shielding and the high-energy model ILU-10 can be used with an X-ray convertor, consisting of tantalum target and water-cooled aluminum collector [40]. The characteristics of the current ILU accelerators are given in Table 3.1

3.2.1.1.3 Low-Energy Electron Accelerators

Electron accelerators discussed in Sections 2.1 and 2.2 generate high-energy electrons (10 MeV or even more) and are used mainly to process thick polymeric products, typically up to 20 mm, and for sterilizing medical devices and food processing. However, the vast majority of current industrial irradiation work, such as cross-linking of thin polymeric films and sheets, polymerization, and cross-linking of coatings, is done by low- to moderate-energy accelerators with low electron energies (less than 1.0 MeV). Low-energy accelerators are used, often as reliable computer-controlled sub-systems, in coating lines, printing presses, laminating machines, etc. Their operating parameters, such as electron energy, EB power, irradiation width, and delivered dose rate, can be matched to the demands of the given industrial process. Such machines are often referred to as *electron processors*.

Table 3.1 Characteristics of ILU Accelerators

Characteristic	Accelerator			
	ILU-6	ILU-8[a]	ILU-10	ILU-14[b]
Electron energy (MeV)	1.2–2.5	0.6–1	2.5–5	7.5–10
Maximum Beam Power (kW)	20	25	50	100
Maximum beam current (mA)	20	30	15	–
Power consumption (kW)	100	80	150	–
Accelerator weight (tons)	2.2	0.6	2.9	–

[a]*With local shielding (thickness in walls 330 mm and in the top 240 mm).*
[b]*In development (at the time of writing).*

The three most common designs of industrial low-energy electron accelerators are shown in Figure 3.12.

The selection of a suitable electron processor type is dictated by the process parameters. *Scanning*-type electron accelerators use a typical small-diameter electron gun, which generates and shapes a pencil beam. The beam is deflected over the window area by using periodically changing magnetic fields. Carefully tuned deflection produces a very uniform beam current (or dose rate) distribution across the beam exit. Energies up to 1.0 MeV can be reached at medium- to high-beam current. Due to limited deflection angle, the length of the scanning horn must be somewhat larger than the length of the exit window.

Single filament design, consisting of a long linear filament to determine the beam current and beam width, eliminates the need for a beam scanning system. Since the current extractable from a unit length of the single filament is limited, multifilament cathode configurations are often used. This also improves the uniformity of the beam current distribution, which is usually less homogeneous than that obtained in scanning horn processors [41, p. 139]. The beam width is limited by the necessity to span the filament across the full width of the exit window to a maximum of about 2 m (80 in). This design is no longer used for currently made machines.

Multifilament design, consisting of an assembly of short, about 20 cm (8 in) long, filaments set parallel to the web direction eliminates the problem of filament support and adjustment of the single filament design. The filaments are mounted between two rigid bars determining the length of the

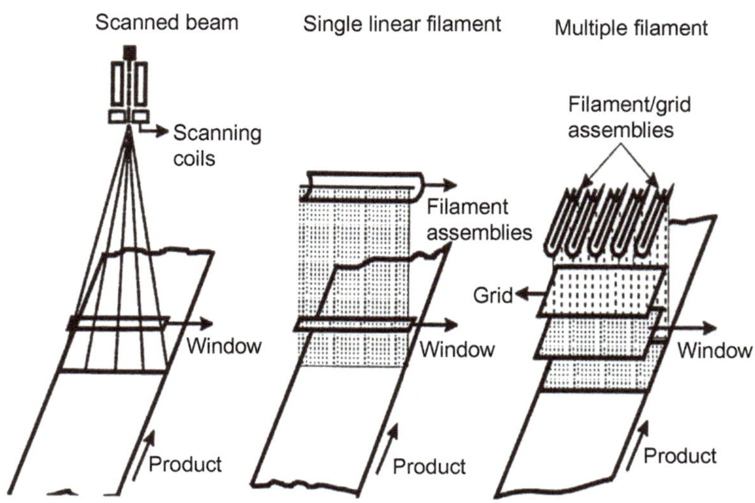

Figure 3.12 Common Designs of Industrial Low-Energy Accelerators.

filament assembly. A control and screen grid define the electron optical extraction conditions for each filament into a single-gap accelerator. The overlapped beams emitted from adjacent cathode filaments form a uniform electron cloud before the acceleration takes place. This design is capable of generating high beam power, dose uniformity, and a wider beam than the single filament design [41].

Single-Stage Scanned Beam Accelerator The accelerator for the EB processor produced by Electron Crosslinking AB in Halmstadt, Sweden, originally developed by Polymer Physik in Tübingen, Germany, uses a classical triode system as electron gun, a single electrode gap, a beam focusing, and deflection system. The acceleration voltage depends on the application and ranges from 150 to 300 kV [42].

The electron gun consists of a spiral-shaped tungsten cathode and a Wehnelt cylinder. These two components not only constitute the electrodes of the acceleration gap but also form the optical assembly to control and shape the EB. Current signals are linear and have repetition frequency of about 800 Hz. They are used to deflect the EB horizontally and vertically over the exit window plane. The scanner may be equipped with two cathodes for maximum output. Then, the width of the exit window is more than double that of a standard unit with a single cathode. The exit window containing the 12−15 μm thick titanium foil is relatively large to ensure effective cooling of the foil. The processor is described in Figure 3.13.

Linear Cathode Electron Accelerators A linear cathode accelerator employs a cylindrical vacuum chamber in which a longitudinal heated tungsten filament cathode is raised to the negative accelerating potential. The electrons emitted from the cathode are accelerated in a single step to the exit window. The exit window as a part of the vacuum chamber is kept at ground potential. The accelerated electrons penetrate the thin titanium window and enter the process zone. A linear cathode processor using a single filament has a dose-speed capacity of about 450 m/min (1500 ft/min) at 10 kGy. In the actual commercial equipment, the cathode assembly contains up to four filaments which is capable of providing a dose-speed capacity of up to 1350 m/min (4500 ft/min) at 10 kGy or a beam current of several hundred mA [41, p. 150]. The processor is supplied with a protective shielding against X-rays in the form of a lead slab. The X-ray shielding and product handling system are designed to match the specific process demands. An example of a linear cathode accelerator is shown in Figure 3.14.

The process control of self-shielding linear cathode processors is provided by programmable process controllers (PLC), which are flexible and upgradable. Addition of man−machine interfaces (MMI) based on personal computers and common software systems allows not only system control and

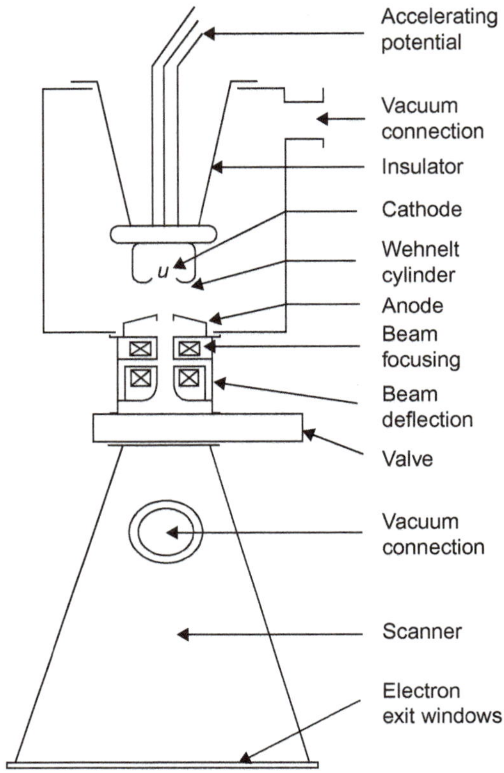

Figure 3.13 Schematic of the Single-Stage Scanned Beam Accelerator. *(Courtesy of Elektron Crosslinking AB.)*

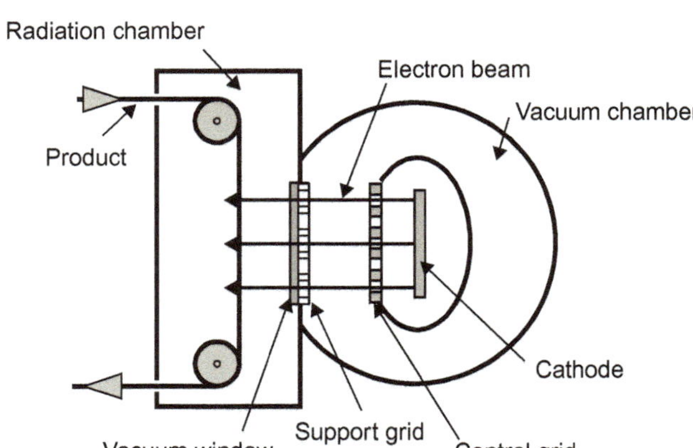

Figure 3.14 Schematic of a Linear Cathode Accelerator. *(Courtesy of Elektron Crosslinking AB.)*

interlocks but also a history operation, communication, and archiving [43]. A series of processors of this design was manufactured by Energy Sciences Inc. (ESI) in Wilmington, MA, USA, under the brand Electrocurtain® in the 1970s and 1980s. Currently the processors made by ESI are designed with multifilament emitters. An example of a self-shielding processor is shown in Figure 3.15.

Another low-energy processor can be described as a vacuum diode operating in the current saturation mode, i.e., the cathode is a directly heated tungsten filament without a control grid. The electrons emitted from the cathode are shaped into EB bundles by a forming electrode and accelerated to the exit window. This diode configuration ensures that practically all electrons emitted are accelerated. Cathode heating can be accomplished through the high-voltage cable. This allows moderate electron currents at low heating power [44]. An example of such a processor was EBOGEN produced by igm Robotersysteme AG in Munich, Germany (Figure 3.16). At the time of writing, the equipment has been reportedly modified and is being produced by Steigerwald Strahltechnik in Maisach, Germany, under the trade name EBOCAM®. It is used mainly for welding and drilling metals.

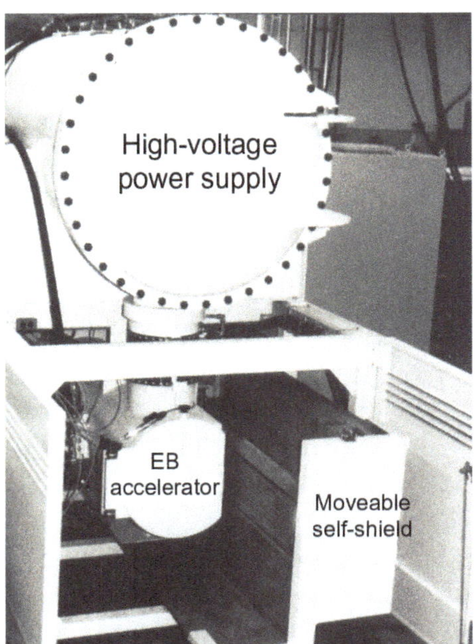

Figure 3.15 Example of a Self-Shielding Processor. *(Courtesy of Energy Sciences Inc.)*

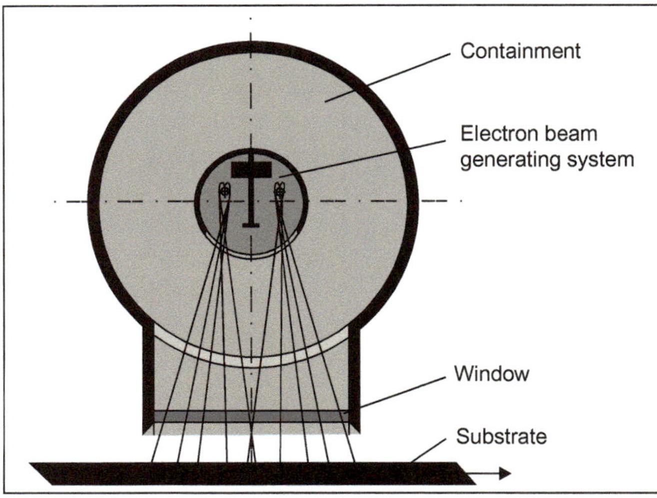

Figure 3.16 Schematic of the EBOGEN Processing Unit. *(Courtesy of igm Robotersysteme AG.)*

Multifilament Linear Cathode Electron Accelerators In this design, a multiple emitter assembly with heated tungsten filaments placed in parallel to the product's direction is used. The beam current is controlled by molybdenum grids held at a common potential. A planar screen grid is placed below the control grids. In the field-free region between the grids, a highly uniform electron density is generated prior to acceleration. The acceleration of electrons occurs between the planar screen grid and the grounded window. This design is represented by the BROADBEAM™ processor that was manufactured by the now defunct RPC Industries in Hayward, CA, USA. Currently, this type of equipment is produced by PCT Engineered Systems in Davenport, IA, USA. The detail of the multifilament array of an EB accelerator is shown in Figure 3.17.

The operating characteristics of the four most widely used low-energy EB processors are given in Table 3.2.

A new window design (cooling and support structure) combined with a multifilament cathode arrangement lead to the development of low-energy, low-cost electron processors, such as Electrocure™ and EZ Cure™ manufactured by Energy Sciences Inc., Wilmington, MA, USA. For example, a 150 kV processor is capable of a dose-speed capacity of 1200 m/min (3950 ft/min) at 10 kGy. An even more powerful 125 kV, 200 kW processor is capable of curing coatings on a web 1650 mm (66 in.) wide with a dose-speed capacity of 2250 m/min (7400 ft/min) at 10 kGy [41, p. 155]. Smaller, lower voltage processors operating between 90 and 110 kV accelerating voltages are capable of speeds up to 400 m/min (1310 ft/min).

Figure 3.17 Multifilament Array of an EB Accelerator. *(Courtesy of M.R. Cleland.)*

Table 3.2 General Operating Characteristics of Different Designs of Selected Commercial Low-Energy Industrial EB Processors

	Accelerator		
Characteristic	**Single Filament**	**Multifilament**	**Single Stage Scanned Beam**
Acceleration voltage (kV)	150–300	150–300	150–300
Working width (mm)	150–2000	300–2500	150–2000
Maximum web speed at 10 kGy, maximum (m/min)	1350	1500	1350
Typical dose variation across the web (%)	±10	±8	±4

Industrial requirements for compact, low-energy EB processing equipment have driven the development of a self-shielded Dynamitron® system (Figures 3.18 and 3.19). The system Easy-e-Beam™ developed by IBA-RDI in Edgewood, NY, is installed in a small footprint enclosure with sufficient shielding for safe operation in an uncontrolled area. The equipment is

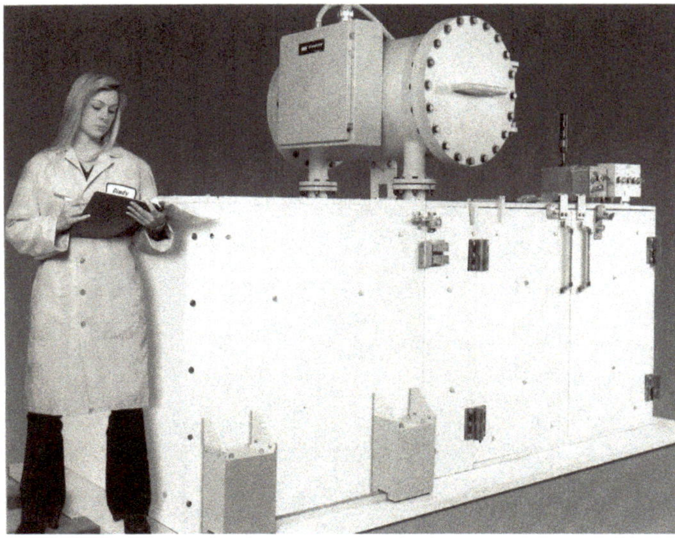

Figure 3.18 Easy-e-Beam™ (Self-Shielding Dynamitron, Overall View). *(Courtesy of IBA Industrial.)*

Figure 3.19 Easy-e-Beam™ (Self-Shielding Dynamitron, High-Voltage Generator and Acceleration Tube). *(Courtesy of IBA Industrial.)*

available with electron energies from 300 to 1000 kV and EB power up to 50 kW [45].

NHV Corporation of Japan also offers machines with varied designs (both scanning and linear cathode) and performance levels under the name Electron Beam Processing System (EPS). The acceleration voltages in these machines are in the range 250–1000 kV with beam power levels in excess of 65 kW. The maximum beam width is 1800 mm (71 in.) [46]. Examples of two types of NHV machines are shown in Figures 3.20 and 3.21.

ELV Accelerators ELV accelerators are developed and manufactured at the Budker Institute of Nuclear Physics. They are low-energy equipment with electron energy ranging from 0.2 to 2.5 MeV, beam current up to 400 mA, and maximum power of 400 kW. The source of high voltage is a cascade generator with a parallel inductive coupling. The rectifier is installed inside the primary winding.

The equipment is used for cross-linking PE insulation in the wire and cable industry; for the production of heat-shrinkable tubing; for the manufacture of synthetic leather, gels and prepregs, and heat-resistant plastic pipes; and for cross-linking of films, rubber, coatings, paints, and lacquers [47].

3.2.2 Commercial EB Reactors—Their Performance and Use

Elektron Crosslinking AB, Halmstad, Sweden

EC-Scanner (Figure 3.22)
Accelerating voltage: 75–300 kV
Electron current: 0–200 mA
Maximum working width: 2 m
Throughput: 9000 kGy m/min
Distribution of dose across over the working width: $< \pm 4\%$.

EC-Beam (Figure 3.23)
Accelerating voltage: 75–250 kV
Electron current: 0–2000 mA
Working width: 0.4–3 m
Throughput: 14,000 kGy m/min
Distribution of dose across over the working width: $< \pm 10\%$.

EC-Print (Figure 3.24)
Accelerating voltage: 75–110 kV
Electron current: 0–200 mA
Working width: 0.1–0.6 m
Throughput: 9000 kGy m/min

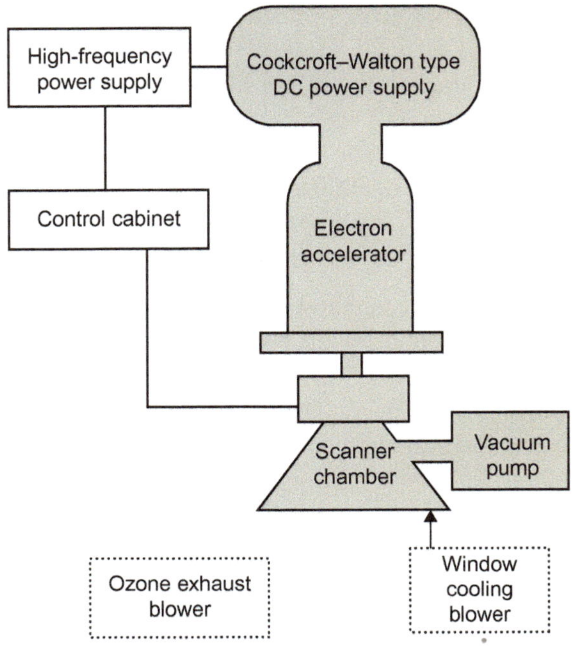

Figure 3.20 Schematic of a Cockroft–Walton EPS Accelerator with Scanner. *(Courtesy of NHV Corporation.)*

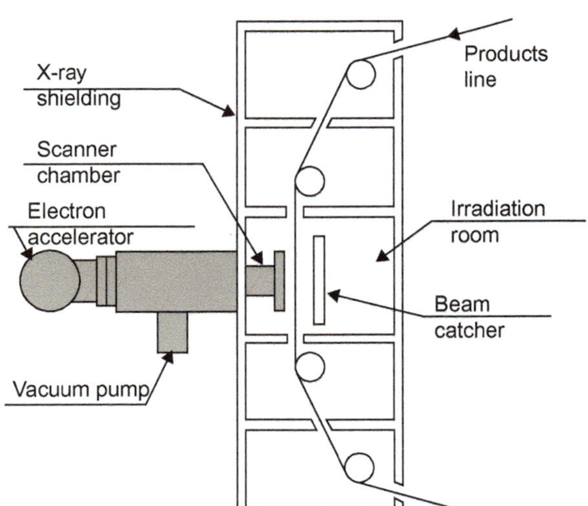

Figure 3.21 Schematic of a Self-Shielding EPS Accelerator for Film Processing. *(Courtesy of NHV Corporation.)*

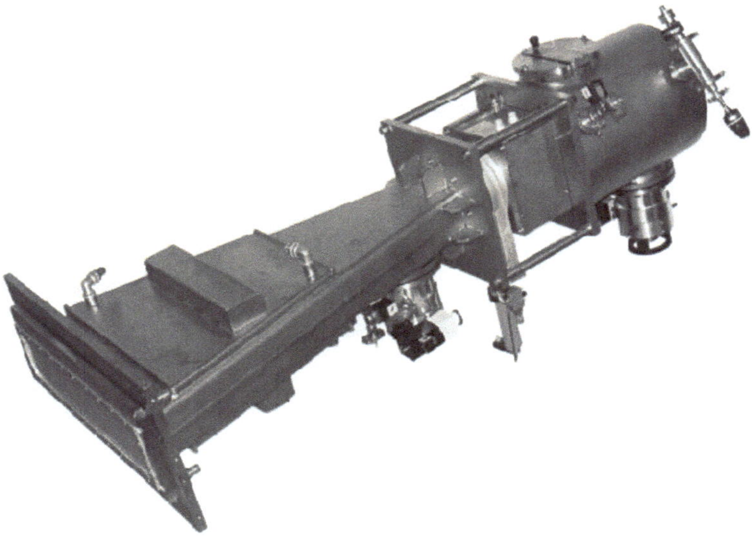

Figure 3.22 EC-Scanner.

Figure 3.23 EC-Beam.

Distribution of dose across over the working width: <± 7%.

EC-Tube (Figure 3.25)

Accelerating voltage: 70–140 kV
Electron current: 0–2 mA
Maximum working diameter: 80 mm
Typical bottle size (material): 3.5 l (PET)

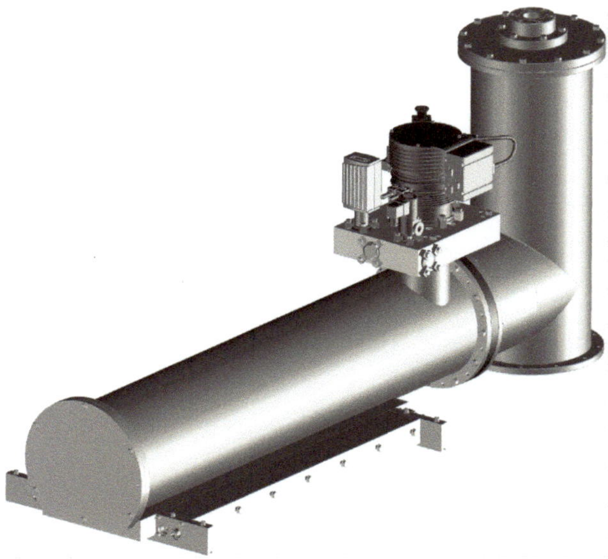

Figure 3.24 EC-Print.

Figure 3.25 EC-Tube.

Throughput: 1 bottle/s
Sterility assurance level (SAL): 10^{-6}.

Energy Sciences Inc., Wilmington, Massachusetts

EZCure-DF (Figure 3.26)
Web width: 20–54 in (0.5–1.35 m)
Speed capability: 1300 ft/min (400 m/min)
Maximum curing thickness: 1.8 mils (45 μm)
Applications: offset printing lines, flexo printing lines, coatings, laminating adhesives for flexible packaging.

EZCure-F-1 (Figure 3.27)
Web width: 20–66 in (0.5–1.675 m)
Speed capability: 1300 ft/min (400 m/min)

Figure 3.26 EZCure-DF.

Figure 3.27 EZCure-F-1.

Maximum curing thickness: 1.8 mils (45 μm)

Applications: flexo printing lines, coatings, laminating adhesives for flexible packaging, film cross-linking.

EZCure-CR (Figure 3.28)

Web width: 20–54 in (0.5–1.35 m)

Speed capability: 1200 ft/min (360 m/min)

Maximum curing thickness: 1.6 mils (40 μm)

Applications: offset printing lines, flexo printing lines, coatings, laminating adhesives for flexible packaging.

EZCure-LS (Figure 3.29)

Web width: 20–54 in (0.5–1.35 m)

Speed capability: 660 ft/min (200 m/min)

Maximum curing thickness: 0.8 mils (20 μm)

Applications: flexo printing lines, coatings for flexible packaging.

Electrocure Casette (Figure 3.30)

Web width: 20–48 in (0.5–1.22 m)

Speed capability: 1300 ft/min (400 m/min)

Maximum curing thickness: 3.0 mils (75 μm)

Applications: offset printing, coatings, laminating adhesives, film cross-linking.

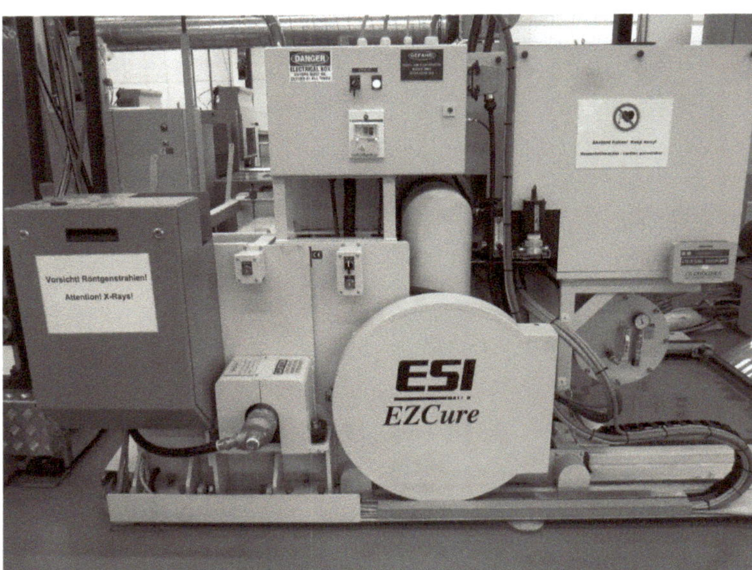

Figure 3.28 EZCure-CR.

Figure 3.29 EZCure-LS.

Figure 3.30 Electrocure Cassette.

Custom Systems (Figure 3.31)
Web width: 20–120 in (0.50–3.05 m)
Speed capability up to 1300 ft/min (400 m/min)
Maximum curing thickness: 16 mils (40 μm)

Figure 3.31 (A and B) Custom Systems.

Designed for maximum depth of cure capability of many substrates, film, and sheet cross-linking, cross-linking of adhesives.

CHV—Compact High-Voltage System (Figure 3.32)
Web width: 18–66 in (0.45–1.65 m)
Speed capability: 1000 ft/min (300 m/min)
Maximum curing thickness depends on system, material, and application
Designed to deliver voltages of 100, 150, 250, and 300 kV
Suitable for product development, application development, pilot runs, and for production runs.

IBA Industrial, Louvain-la-Neuve, Belgium

Dynamitron—5 Standard Models (Figure 3.33A and B)

1. Rated energy: 560 keV
   Rated beam current: up to 160 mA
   Minimum operating voltage: 300 kV
   Control system: PLC Siemens S7
2. Rated energy: 800 keV
   Rated beam current: up to 160 mA
   Minimum operating voltage: 400 kV
   Control system: PLC Siemens S7

Figure 3.32 CHV-Compact High-Voltage System.

Figure 3.33 (A) Dynamitron Assembly, 5 MeV, 300 kW. (B) Dynamitron Rectifier Column, 3.0 MeV, 150 kW. *((A) Courtesy of M. R. Cleland. (B) Courtesy of IBA Industrial.)*

3. Rated energy: 1.5 MeV
   Rated beam current: up to 65 mA
   Minimum operating voltage: 500 kV
   Control system: PLC Siemens S7

4. Rated energy: 3 MeV
   Rated beam current: up to 50 mA
   Minimum operating voltage: 1100 kV
   Control system: PLC Siemens S7
5. Rated energy: 5 MeV
   Rated beam current: up to 30 mA
   Minimum operating voltage: 1700 kV
   Control system: PLC Siemens S7

Main configurations:

- Self-shielded (available up to 800 keV)
- Right angle (limited shielding, the vessel does not require shielding)
- In-line (shielding of the vessel and the irradiation area is required)

Applications: Polymer cross-linking (wire and cables, rubber and tires, films and sheets, composites, heat-shrink systems), polymerization, sterilization of medical devices, aseptic packages semiconductors, and others.

Rhodotron—4 Standard Models (Figure 3.34)

1. Rhodotron TT 100
   Energy: 2.5–10 MeV
   Power range: up to 40 kW

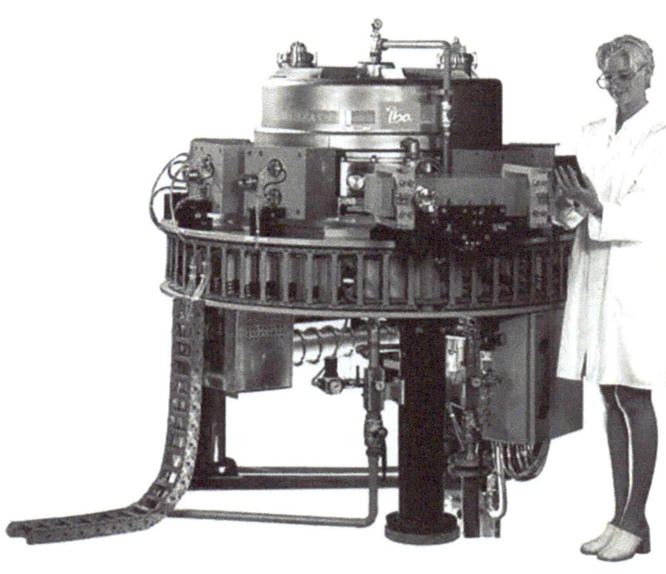

Figure 3.34 Rhodotron TT 100.

Diameter: 1.6 m
Height: 1.7 m
MeV per pass: 0.833
Number of passes: 12

2. Rhodotron TT 200
Energy: 2–10 MeV
Power range: 40 to 100 kW
Diameter: 3 m
Height: 2.4 m
MeV per pass: 1
Number of passes: 10

3. Rhodotron TT 300 (Figure 3.35)
Energy: 2–10 MeV
Power range: 40–420 kW
Diameter: 3 m
Height: 2.4 m
MeV per pass: 1
Number of passes: 10
Note: This model can be used for both EB and X-ray applications.

Figure 3.35 Rhodotron TT 200 and TT 300.

4. Rhodotron TT 1000 (Figure 3.36)
   Energy: 5–10 MeV
   Power range: 100–560 kW
   Diameter: 3 m
   Height: 3.4 m
   MeV per pass: 1.166
   Number of passes: 6
   Note: This model is specially designed for X-ray applications.

Applications:

- Polymer cross-linking
- Polymer-based composite curing
- Sterilization of medical devices
- Food irradiation

Easy-e-Beam System (Figure 3.37)
   Beam voltage: 800 kV
   Beam current: 70–100 mA

Figure 3.36 Rhodotron TT 1000.

Figure 3.37 Easy-e-Beam System™.

Scan width: 0.91 m (36 in)
Maximum system processing speed: 1000 m/min (3200 ft/min)
Typical product range: thin wall wires from 24 to 9 G (0.22 to 6 mm^2 cross-sectional area)
Overall dimensions: 10.3 m × 5.3 m × 5.3 m (L × W × H)
34 ft × 17 ft × 17 ft (L × W × H)
Weight: 50 metric tons (110,000 lb).

The system includes EB accelerator, wire and cable handling system, integrated safety shielding, PLC control system, vacuum pumping system, radio frequency oscillator, insulating gas subsystem, and water cooling system.

Application: Wire and cable production (see also Section 6.1.1.3) (Figures 3.38 and 3.39).

NHV Corporation, Kyoto, Japan

Specifications and ratings of high- and medium-energy processors are given in Tables 3.3 and 3.4 [46].

PCT Engineered Systems, LLC, Davenport, Iowa

BroadBeam LE Series (Figure 3.40)
Operating voltage: 70–150 kV
Standard web widths: 36 in (0.91 m), 54 in (1.37 m), 72 in (1.80 m)
Surface dose rate: 12,000 kGy m/min

Figure 3.38 High-Energy EPS Processor with Scanner.

Figure 3.39 Self-Shielding EPS Medium-Energy Processor.

Table 3.3 High-Energy EPS Series, Specifications, and Ratings

Type		EPS-1500	EPS-2000	EPS-3000	EPS-5000
Acceleration voltage (kV)		1500	2000	3000	5000
Beam current (mA)		65	50	30	30
Irradiation width (mm)		1200	1200	1200	1200
Mass throughput rate tons/hour at	20 kGy	9.4	9.5	8.8	14.6
	50 kGy	3.7	3.8	3.5	5.8
	100 kGy	1.9	1.9	1.8	2.9
Size (mm)	Width	4000	2900	3000	3800
	Height	4350	6950	8950	11200

Source: *EPS Technical Guide, NHV Corporation, Kyoto.*

Cross-web dose uniformity: ±8% guaranteed
Integrated chill roll
Low-profile, side-fire orientation
Application: in printing, coil coating, and specialty cross-linking applications.

BroadBeam EP Series (Figure 3.41)
Operating voltage: 125–300 kV
Standard web widths: up to 130 in (3.3 m)
Throughput: up to 15,000 kGy m/min
Line speeds: up to 2000 fpm (600 m/min)
Cross-web dose uniformity: ±8% guaranteed
Beam orientation: horizontal, vertical, and angled
Applications: coil coating, film cross-linking, curing silicone release coatings, curing laminating adhesives for paper, foil, and wood products.

BroadBeam OSD Series (Figure 3.42)
Operating voltage: 90–125 kV
Standard web widths: 52 in (1.32 m), 36 in (0.91 m)
Throughput: 30 kGy at 400 m/min; 12,000 kGy m/min
Cross-web dose uniformity: ± 8% guaranteed
Integrated chill roll
Beam direction: side-fire
Suitable mainly for printing and coating applications

Table 3.4 Medium-Energy EPS Series, Specifications, and Ratings

Type		EPS-500			EPS-800		EPS-1000	
Accelerator voltage (kV)		500	500	500	800	800	1000	1000
Beam current (mA)		65	100	150	65	100	65	100
Irradiation width (mm)		1200	1800	1800	1200	1800	1200	1800
Irradiation speed (m^2/min), at dose	20 kGy	52	80	120	43	66	36	56
	50 kGy	21	32	48	17	26	14.5	22
	100 kGy	10	16	24	9	13	7	11
Size (mm)	Width	3800	3800	3800	3800	3800	4000	4000
	Height	3050	3500	3500	3350	3750	3700	4100

Source: *EPS Technical Guide, NHV Corporation, Kyoto.*

Figure 3.40 BroadBeam LE.

Figure 3.41 BroadBeam EP.

Figure 3.42 BroadBeam OSD.

3.2.3 Laboratory EB Units

Several companies have developed equipment for laboratory work intended mainly for material and process research and development. Examples are listed below:

Comet AG:

e-beam Test Lab (Figure 3.43)
Accelerating voltage: 70–210 kV
Maximum beam power: 2.25 kW
Maximum beam current: 20 mA (up to 112 kV)
Dose uniformity: ±10%
Sample transport speed: 3–30 m/min
Sample size (A4 format): $210 \times 297\ mm^2$
Height: adjustable up to 50 mm
Nitrogen inerting.

Elektron Crosslinking AB:

EC-LAB 400 (Figure 3.44)
Accelerating voltage: 80–300 kV
Electron current: 0–30 mA
Maximum working width: 0.4 m
Web speed: 0–150 m/min

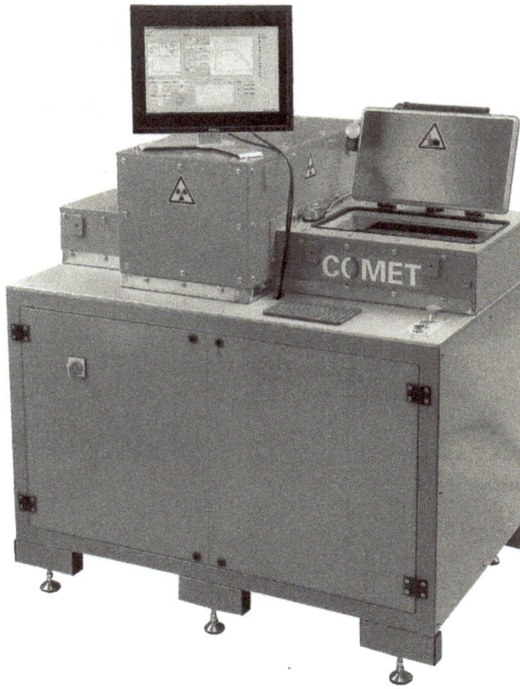

Figure 3.43 Comet Laboratory System.

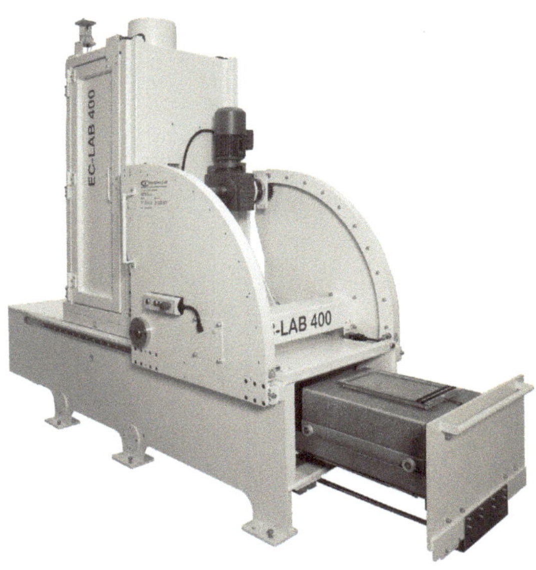

Figure 3.44 EC-LAB 400.

Throughput: 4500 kGy m/min
Cassette speed: 5–30 m/min

Energy Sciences Inc.:

EZLab (Figure 3.45)
Acceleration voltage (range): 80–100 kV
Beam current: 1–5 mA
Irradiation width: 150 mm (6 in)
Sample size: 150 × 150 mm^2 or continuous 150 mm wide web
Irradiation gap: 10–15 mm
Nitrogen purged system.

Ushio America:

U-electron™ (Figure 3.46)
Irradiation gap: 1–50 mm

Figure 3.45 EZLab.

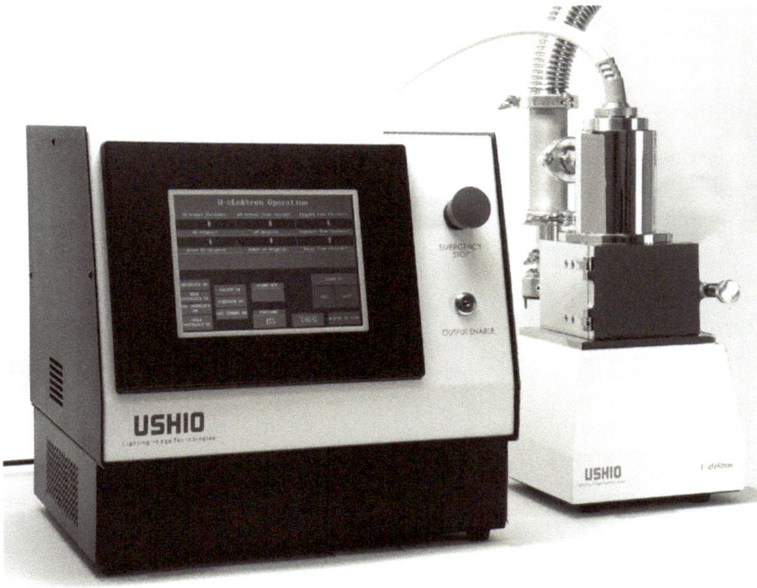

Figure 3.46 U-Electron™.

Accelerator potential: 10–50 keV
Cathode current: 0–200 μA
Gun vacuum: 1×10^6 Torr
Chamber atmosphere: ambient or purge connections
Applications: R&D work

3.3 X-Ray Processing Equipment

As pointed out in Section 1.2.2, one way to produce high-energy X-rays is utilizing bremsstrahlung, i.e., by conversion of high-energy electrons into X-ray photons. The high-energy X-rays for industrial applications are produced by interposing a metal target between the EB and the product to be treated by the X-radiation. To enhance the conversion of electron to photon, these X-ray targets are made of metals with high atomic (Z) number [48–50]. Water-cooled tantalum is preferred for large area targets. The conversion efficiency of this process depends on accelerator energy and the atomic number of the target; efficiencies ranging between 8% and 12% have been shown for electron energies between 5.0 and 7.5 MeV [51]. X-radiation has a forward-peaked emission and the rate at which a material receives the X-ray photons, the dose rate, can be controlled by a combination of the distance from the target, the beam current, and under-beam transport speed. Because

Table 3.5 X-Ray Processing Throughput Potential

Source	Penetration in Unit Density[a], mm	Emitted Power (kW)	Potential Throughput (kg/h)
10 MeV linac, EB mode	38	20	720
10 MeV Rhodotron™ EB mode	38	200	7200
10 MeV linac, 20 kW X-ray mode	480	3.2	109
7 MeV Rhodotron 700 kW, X-ray mode	450	77	2772
5 MeV Dynamitron™ 300 kW, X-ray mode	385	24	864

[a]*Equal entrance, equal exit.*
Note: *Assuming an under beam process efficiency of 25% (typical of cart system), and a dose of 25 kGy and X-ray conversion efficiencies of 16% at 10 MeV, 11% at 7 MeV, and 8% at 5 MeV.*
Source: *Courtesy of A.J. Berejka.*

of low efficiency of the conversion, X-ray irradiation has been made commercially feasible only with the development of high-energy and high-beam-power electron accelerators, typically 300 kW at 5.0 MeV or 700 W at 7.0 MeV. X-ray penetration is much greater than that of the highest energy industrial EB systems and is even better than γ-ray penetration. X-ray dose rates are at least one order of magnitude higher than those of γ-rays but significantly less than those of EBs (see Table 1.2). Product throughput rates for X-ray processing based on high-power EB equipment can exceed that of high electron energy, low- to medium-power 10 MeV linacs that are used in industrial applications as seen in Table 3.5.

Despite the high-power and high-dose rate, EB has a serious limitation in the penetration of thick products when compared to X-rays. However, if the EB is converted into X-rays, it overcomes the problem of low penetration. Since the conversion process to X-rays has a low efficiency, thus X-ray irradiation has been made commercially feasible only with the development of high-energy and high-beam-power electron accelerators. An example of an X-ray target is shown in Figure 3.47A. An irradiation room with EB and X-ray ports is shown in Figure 3.47B.

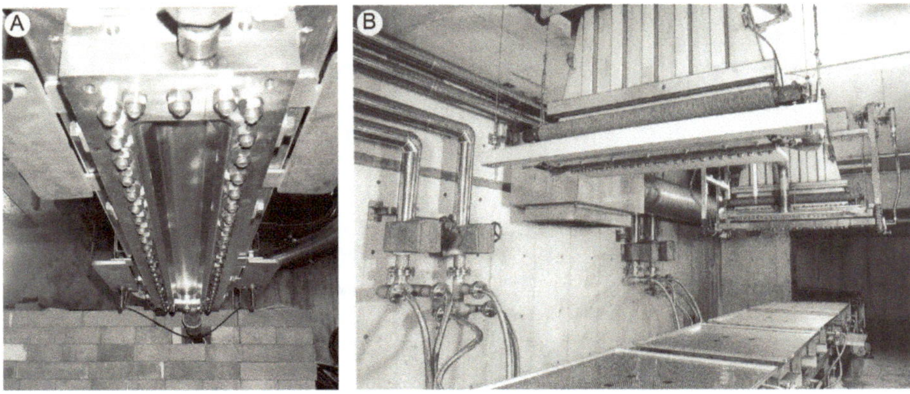

Figure 3.47 (A) Example of an X-ray Target and (B) Irradiation Room with EB and X-ray Ports. *(Courtesy of M. R. Cleland.)*

3.4 Current Developments in EB Processing Equipment and Technology

Current developments in the EB curing technology include:

- Lowering equipment cost
- Reducing size and weight of the curing equipment
- Elimination or reduction of the use of nitrogen inerting.

Lowering accelerator voltage is the single major factor helping to meet the above requirements. Lower accelerating voltage means a less expensive power supply, smaller size accelerator head, and less X-ray shielding [41, p. 156]. For example, a newly designed 75 kV MicroBeam-LV processor (Energy Sciences, Inc.) is capable of curing coatings up to 0.001 in (25 μm) thick (at density 1). Complete cure is achieved at doses of 1–2.5 Mrad

Table 3.6 Comparison of Standard and Low-Voltage EB Processors for a 48 in Wide Web

Dimensions	Standard Processor	Low-Voltage Processor
In-line dimensions, ft (mm)	7 (2135)	4.5 (1370)
Height, ft (mm)	10 (3050)	4.5 (1370)
Depth, ft (mm)	12 (3660)	6 (1830)

(10−25 kGy), which is considerably lower than the typical 3 Mrad (30 kGy) required for the 150 kV equipment. Compared to a standard processor, the total volume is 50% smaller. Comparison of standard and low-voltage processors, both for 48 in wide web, is given in Table 3.5 [52]. PCT Engineered Systems is offering BroadBeam LE Series with accelerating voltages in the range from 70 to 125 kV, to be used in coil coating, web offset presses, flexo presses, laminating presses, and specialty cross-linking with throughput typically 30 kGy at 400 m/min (1312 ft/min) (Table 3.6).

The penetration depth of electrons into the substrate decreases at lower energy. Also, the surface dose per mA of beam current increases and thin coatings and inks can be cured with higher efficiency and speed. The higher surface dose produces a higher concentration of radicals, which in turn reduces the oxygen inhibition [41].

Model P-200, developed by ESI, is a low-energy processor designed to deliver voltage between 100 and 200 kV, with a speed capability up to 400 ft/min (122 m/min) and maximum curing thickness of 6.0 mil (150 μm) at unit specific gravity and maximum web thickness of 48 in (1.22 m). Model P-200 is suitable for small-scale production and product/process development, which is shown in Figure 3.48.

Relatively recent developments are electron generators that are essentially sealed vacuum tubes with a 2.5 μm thick silica ceramic window or a 6 μm titanium window as the beam exit. Such EB tubes or bulbs are capable of operating at 50−150 kV [41,53−57] and can be connected into modules to

Figure 3.48 P-200 Processor. (*Courtesy of Energy Sciences, Inc.*)

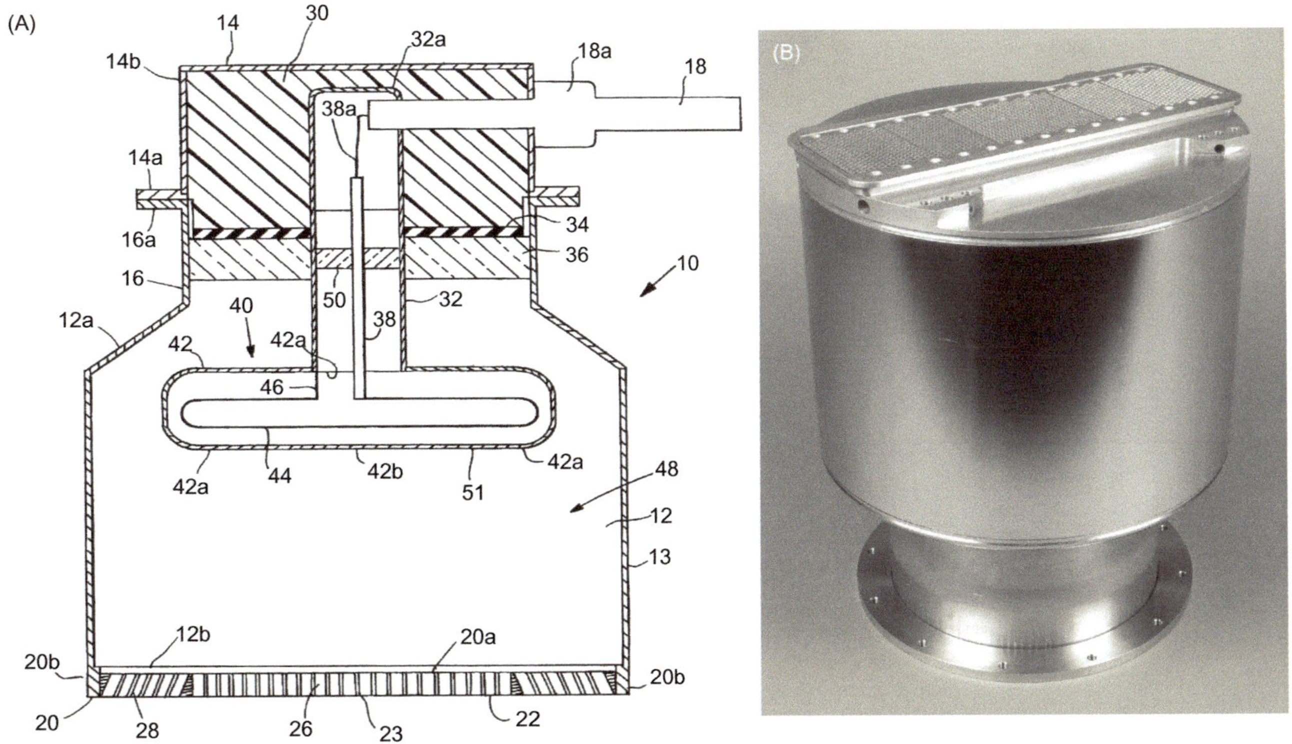

Figure 3.49 (A) Patented AEB Emitter. (B) 25 inch (635 mm) AEB Emitter.

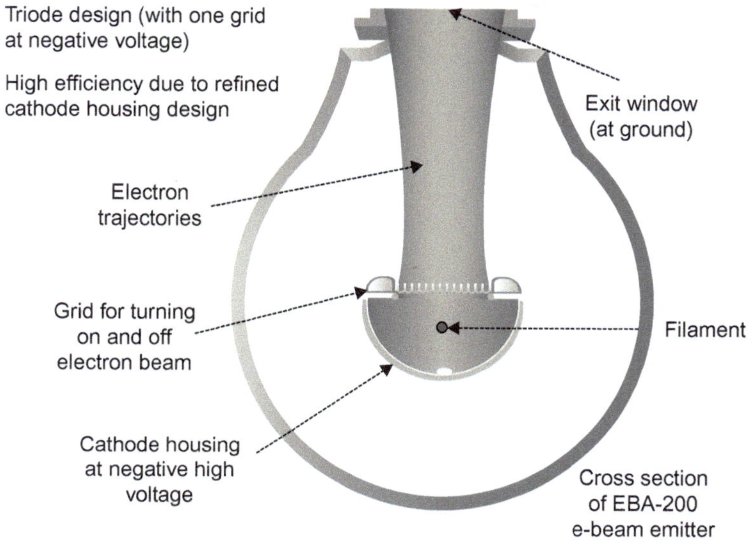

Figure 3.50 Schematic of the Comet e-Beam Emitter EBA-200. *(Courtesy of Comet AG.)*

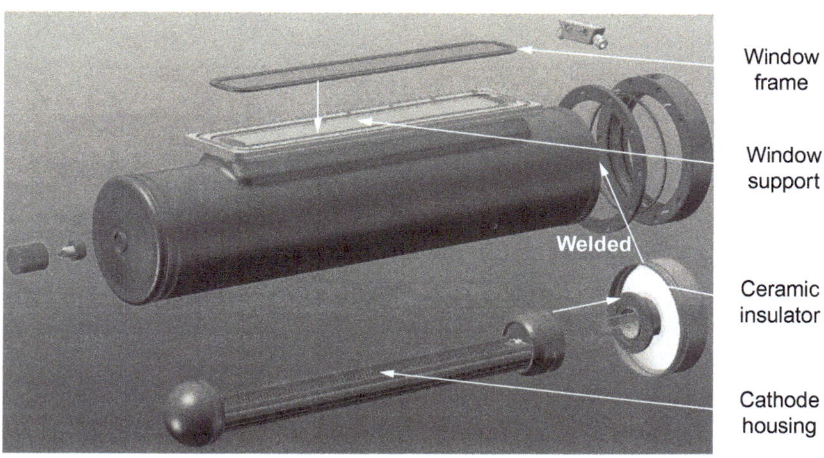

Figure 3.51 Modular Design of Comet e-Beam Emitter EBA-200. *(Courtesy of Comet AG.)*

cure wide webs. It is conceivable that miniature EB tubes may provide efficient low cost means to cure thin coatings, inks, and adhesives thus offering an alternative to irradiation by UV lamps. A schematic of one of the patented emitters developed by now defunct AEB Inc. is shown in Figure 3.49A and an actual 10 in emitter is shown in Figure 3.49B. Another sealed beam emitter was developed by Comet AG in Switzerland, which has a triode design with one grid at negative voltage (Figure 3.50). The emitter is based on the 450 kV bipolar X-ray tube design, produced by that company [58]. The modular design is shown in Figure 3.51.

References

[1] Gamma Irradiators for Radiation Processing. International Atomic Energy Agency. Vienna, Austria. p. 14.
[2] Dosimetry for Food Irradiation. International Atomic Energy Agency. TRS No. 409. Vienna, Austria; 2002.
[3] Manual on Panoramic Gamma Irradiators, Categories I and III. International Atomic Energy Agency. IAEA-PRSM-8. Vienna, Austria; 1996.
[4] Beying A. RadTech Europe'97, Lyon, France: Conference Proceedings; June 16–18, 1997. p. 77.
[5] Läuppi UV. RadTech Europe 97, Lyon, France: Conference Proceedings; 1997. p. 96.
[6] Bradley R. Radiation technology handbook. New York, NY: Marcel Dekker; 1984. p. 37 (1983).
[7] Artandi C, Van Winkle Jr W. Nucleonics 1959;17:86.
[8] Charlton EE, Westendorp WF, Dempster LE. *J Appl Phys* 1939;10:374.
[9] Cleland MR. In: Singh A, Silverman J, editors. Radiation processing of polymers. [chapter 3], Munich: Carl Hanser Verlag; 1992. p. 28.
[10] Sakamoto I, Mizusawa K. Radiat Phys Chem 1981;18:1341.
[11] Scharf W. Particle accelerators and their uses. New York, NY: Harwood Academic Publishers; 1986.
[12] Abramyan EA. Industrial electron accelerators and applications. Washington, DC: Hemisphere Publishing Company; 1988.
[13] Van de Graaff RJ. US patents 3187208 (1965), 3289066 (1966), 3323069 (1967).
[14] Emanuelson R, Fernald R, Schmidt C. Radiat Phys Chem 1979;14:343.
[15] Sakamoto I, et al. *Radiat Phys Chem* 1985;25:911.
[16] Cleland MR. In: Singh A, Silverman J, editors. Radiation processing of polymers. Chapt 3, Munich: Carl Hanser Verlag; 1992. p. 29.
[17] Cleland MR, Thompson CC, Malone HF. Radiat Phys Chem 1977;9:547.
[18] Thompson CC, Cleland MR. Nucl Instrum Methods Phys Res 1989; B40/41:1137.

[19] Morganstern KH. Proceedings of the conference on sterilization of medical products, vol. 4. Montreal: Johnson & Johnson, Polyscience Publications; 1986.
[20] Odera M, Nagakura K, Tanaka Y. Radiat Phys Chem 1990;35:534.
[21] Cleland MR, Thompson CC, Strelczyk M, Sloan DP. Radiat Phys Chem 1990;35:632.
[22] Cleland MR. In: Singh A, Silverman J, editors. Radiation processing of polymers. [chapter 3], Munich: Carl Hanser Verlag; 1992. p. 30.
[23] Haimson J. Proceedings of the conference on sterilization by ionizing radiation 1. Montreal: Johnson & Johnson, Multiscience Publications Ltd; 1974.
[24] Haimson J. IEEE Trans Nucl Sci 1975;NS-22:1303.
[25] Anonymous. Beta-gamma news brief. Beta-Gamma 1989;3:38.
[26] Auslender VL, et al. US patent 4140942; 1979.
[27] Cleland MR. In: Singh A, Silverman J, editors. Radiation processing of polymers. [chapter 3], Munich: Carl Hanser Verlag; 1992. p. 34.
[28] Barletta WA. Beam research program, energy and technology review. Livermore: Lawrence Livermore National Laboratory; 1984.
[29] Humphries Jr S. Principles of charged particles acceleration. New York, NY: John Wiley & Sons; 1986.
[30] Lapostolle PM, Septier AL. Linear accelerators. Amsterdam: North-Holland Publishers; 1970.
[31] Abs M, Jongen Y, Poncelet E, Bol J-L. Rad Phys Chem 2004;71 (1–2):285.
[32] Lancker MV, Herer A, Cleland. MR, Jongen Y, Abs M. Nucl Instrum Meth B 1999;151:242.
[33] Jongen Y, Abs M, Genin F, Nguyen A, Capdevilla JM, Defrise D. Nucl Instrum Meth B 1993;79:8650.
[34] Korenev S. Rad Phys Chem 2004;71:535.
[35] Korenev S. Rad Phys Chem 2009;71:277.
[36] Meissner J, Abs M, Cleland MR, Herer AS, Jongen Y, Kuntz F, Strasser A. Rad Chem Phys 2000;57:647.
[37] Bassaler JM, Capdevilla JM, Gal D, Lainé F, Nguyen A, Nicolai JP, Uniastowski K. Nucl Instrum Meth B 1992;68(1–4):92.
[38] ILU Accelerators. Brochure from the Budker Institute of Nuclear Physics (BINP). Novosibirsk, Russia (no date).
[39] Podobaev VS, Bezuglov VV, Briazgin AA, Chernov KN, Cheskidrov VG. Status of ILU-14 electron accelerator. Proceedings of the RuPAC 2010. Protvino, Russia.

[40] Auslender VL. Industrial electron accelerators type ILU. Proceedings of the RuPAC 2006. Novosibirsk, Russia.
[41] Mehnert R, Pincus A, Janorsky I, Stowe R, Berejka A. UV & EB curing technology & equipment. John Wiley & Sons Ltd., Chichester and SITA Technology Ltd.; 1998.
[42] Holl P. Radiat Phys Chem 1985;25:665.
[43] Meskan DA, Klein FA. Proceedings, RadTech Europe'97. Lyon, France; 1997. p. 114.
[44] Schwab U. Proceedings of the RadTech Europe'97. Lyon, France; 1997. p. 114.
[45] Galloway RA, DeNeuter, Lisanti, TF, Cleland MR. Paper CP 680 at the 17th international conference applications of accelerators in research and industry, 12–16 November 2002, Denton, TX: American Institute of Physics; 2003.
[46] Electron Beam Processing System (EPS) Technical Guide, NHV Corporation, Kyoto, Japan.
[47] Kuksanov NK, et al. High power accelerators for industrial application. Proceedings of the RuPAC-2010. Protvino, Russia (THCHZ020); Bulletin accelerators of the ELV type, status, developments, applications from the Budker Institute of Nuclear Physics (BINP). Novosibirsk, Russia (no date).
[48] Farrell JP. Rad Phys Chem 1979;14(3–6):377–87.
[49] Farrell JP, Seltzer SM, Silverman J. Rad Phys Chem 1983;14 (3–5):469–78.
[50] Seltzer SM, Farrell JP, Silverman J. IEEE Trans Nucl Sci 1983;30(2): 1629–33.
[51] Industrial Electron Beam Processing. Vienna, Austria: International Atomic Energy Agency; 2009. p. 10.
[52] Maguire EF. RadTech Rep 1998;12(5):18 1998.
[53] Davis JI, Wakalopulos G. Proc RadTech North Am'96 1996. p. 317.
[54] Wakalopulos G. RadTech Rep 1998;12(4):18 1998.
[55] Brochure Advanced Electron Beams. Wilmington, MA.
[56] Berejka AJ, et al. Proc RadTech North Am 2009 [Indianapolis, IN] 2002; p. 919.
[57] Avnery T. US patents 5962995 (1999), 6545398 (2003) to advanced electron beams.
[58] Haag W. Sealed electron beam emitter for use in narrow web curing, sterilization and laboratory applications. Paper presented at the RadTech UV & EB 2012, April 30–May 2, Chicago, IL: RadTech International North America.

Recommended Further Reading

Makuuchi K, Cheng S. Radiation processing of polymer materials and its industrial applications. Hoboken, NJ: John Wiley & Sons; 2012.

Industrial Radiation Processing with Electron Beams and X-Rays. International Atomic Energy Agency. Vienna, Austria; 2011.

Gamma Irradiators for Radiation Processing. International Atomic Energy Agency. Vienna, Austria.

Drobny JG. Radiation technology for polymers. 2nd ed. Boca Raton, FL: CRC Press; 2010.

Industrial Electron Beam Processing. International Atomic Energy Agency. Vienna, Austria; 2009.

Koleske JV. Radiation curing of coatings. West Conshohocken: ASTM International; 2002.

Mehnert R, Pincus A, Janorsky I, Stowe R, Berejka A. UV & EB curing technology & equipment. London and Chichester: SITA Technology Ltd./ John Wiley & Sons Ltd.; 1998.

Cleland M. R In: Singh A, Silverman J, editors. Radiation processing of polymers. [chapter 3], Munich: Carl Hanser Verlag; 1992.

Singh A, Silverman J, editors. Radiation processing of polymers. Munich: Carl Hanser Verlag; 1992.

Bradley R. Radiation technology handbook. New York, NY: Marcel Dekker; 1983.

Taniguchi N, Ikeda M, Miyamoto I, Miyazaki T. Energy-beam processing of materials. Oxford: Clarendon Press; 1989.

Seidel JR. In: Randell DR, editor. Radiation curing of polymers. London: Royal Society of Chemistry; 1987.

Bakish R, editor. Introduction to electron beam technology. New York, NY: John Wiley & Sons; 1962.

4 Electron Beam Processes

OUTLINE

4.1 Introduction

Since EB irradiation is the most widely used industrial method for processing polymers and polymeric systems by ionizing radiation, this chapter will focus on these EB processes and describe them in some detail. As pointed out in Section 2.2, electrons, capable of exciting and ionizing molecules, have energies in the range from 5 to 10 eV. They are produced from fast electrons by energy degradation. As they penetrate solids and liquids, they generate ions, radicals, and excited molecules. The ionization results from inelastic collisions of fast electrons with the medium and in the process the electrons lose energy. An empirical relationship expressing the energy of the electrons to their depth of penetration is due to Grun [1]:

$$R_G = 4.57E_0^{1.75} \quad (4.1)$$

where R_G is Grun range in μm and E_0 is the electron energy in keV.

This correlation is valid for a wide variety of materials, including polymers, e.g., polystyrene, and metals (e.g., aluminum). Thus, as the energy of the electrons increases, so does their penetration depth and the amount of energy for high-energy electrons dissipated is small and constant over a large depth [2]. The stopping power of a material, which is the energy loss per unit path of an incident electron, depends on the density of the medium and if it is a multicomponent system, it also depends on the relative concentrations of the individual components and on their molecular weights. This is important

Drobny: Ionizing Radiation and Polymers.
DOI: http://dx.doi.org/10.1016/B978-1-4557-7881-2.00004-3

in cases where the organic medium contains pigments, which slow down the incident electrons without yielding useful species [3]. However, the "slowed-down" electrons react more readily with organic species [4]. The reactive species are dispersed randomly throughout the entire thickness of the material as illustrated in Figure 2.3.

The energy deposited in the irradiated material causes a temperature rise (ΔT), which depends on absorbed dose and specific heat:

$$\Delta T = 0.239D/c \tag{4.2}$$

where D is absorbed dose in kGy and c is the specific heat in cal/g°C of the irradiated material. Examples of temperature rise values for selected materials are given in Table 4.1.

Radiation processing of monomers and polymers by EB, such as polymerization and copolymerization of monomers, cross-linking, grafting, and degradation of polymers, is induced by these different chemically reactive species [5–7].

Somewhat simplified, the process of interaction of high-energy electrons with organic matter can be divided into three primary events [8, p. 41]: ionization, excitation, and capture of electron (see Section 2.2).

Besides these primary reactions, various secondary reactions arise in which ions or excited molecules take part [9]. The final result of these three events is that through the diverse primary and secondary fragmentations, radicals are formed that can initiate a free-radical processes, leading to polymerization, cross-linking, backbone or side-chain scissions, structural rearrangements, and others. The complete and rather eventful cascade of reactions triggered by the primary excitation of molecules may take up to several seconds. Under special conditions, such as when the polymer is in the glassy state or when the reactions occur in a crystalline matrix, transient species may survive for hours or even days. The energy deposited does not always

Table 4.1 Temperature Rise Values for Selected Materials in Dependence on Absorbed Dose

Material	**ΔT (°C/kGy)**
Polyethylene	0.43
Polypropylene	0.52
Polyvinyl chloride	0.75
Aluminum	1.11
Copper	2.63

cause change in the precise location, where it is originally deposited and it can migrate and affect considerably the product yield [9]. Migration of charges is another form of energy transfer in polymers, which can be electronic or ionic in nature. It can be either negative or positive and depends on temperature [10,11]. Migration of radicals has been observed in irradiated polyethylene [12].

Although radical cations are generated in some electron-irradiated monomers (e.g., vinyl ethers or epoxies), efficient cationic polymerization is not observed [13]. Under certain conditions (addition of iodonium, sulfonium, or sulfoxonium salts), cationic polymerization with the use of EB irradiation can be induced [14]. Several studies on radiation cross-linking of elastomers support the concept of ionic mechanism [[15–19] [20], p. 592]. Recently published work reported the use of Irgacure 250 (a diaryliodonium salt) in an EB curing formulation. The major advantage of this system is that inerting during irradiation is not necessary [21].

In curing applications, electrons have to penetrate the reactive solids and liquids with typical masses per unit area of one to several hundred g/m^2 ($1\ g/m^2 = 1\ \mu m$ at unit density). As shown in the depth–dose profiles in Figure 4.1 [9], electrons capable of producing any chemical change must have energies greater than 100 keV. Electrons with such energy can be generated by single-gap, low-energy electron accelerators. *The electron penetration range* is related to the path length which the electron travels during the energy degradation process. It can be estimated from the depth–dose

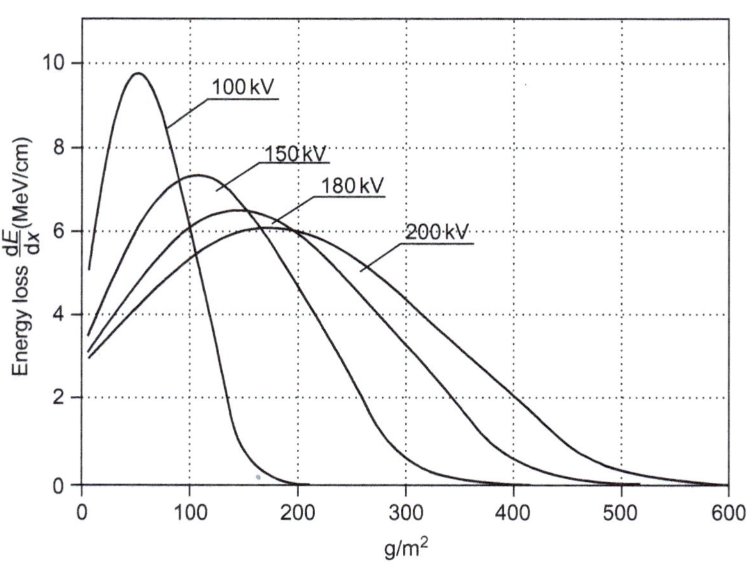

Figure 4.1 Depth of Dose Distribution for Electron Energies 100–200 keV.

distribution. Electron penetration range in g/m^2 as the function of electron energy is shown in Figure 4.2.

The reactions initiated by the formation of radicals are the most important ones in the EB curing process. They can be inhibited by radical scavengers, such as oxygen, which, with its unpaired electrons, is an outstanding electron acceptor and consequently a radical inhibitor. It can readily inhibit polymerization reactions proceeding by the free-radical mechanism typical mainly for acrylates and other vinyl monomers. This process is particularly pronounced on the surface of a coating that is being cured by EB. Therefore, in practical curing applications, nitrogen blanketing is used to prevent direct contact of oxygen with the surface of the material being treated by the radiation [8, p. 42].

The energy levels of industrial accelerators are in the range from 75 keV to 10 MeV. Lower energy accelerators would lose a great portion of their beam power in the beam window and in air; higher energy levels involve the risk of induced radioactivity. For mid-energy (500 keV to 5 MeV) and high-energy (5 to 10 MeV) electron accelerators, it is common to express beam penetration on the basis of *equal-entrance*, *equal-exit* exposure in the unit

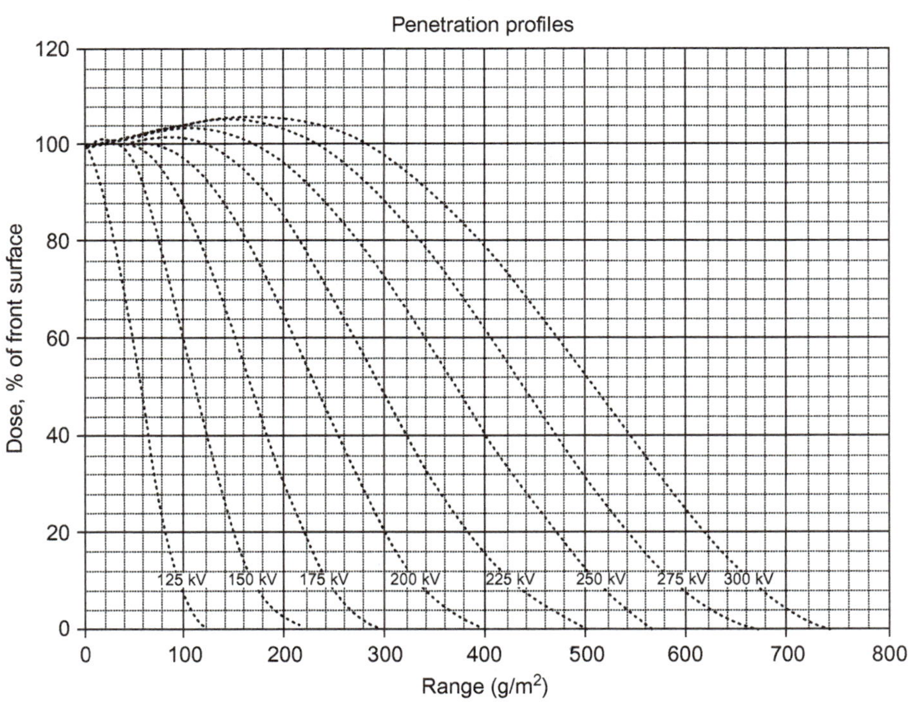

Figure 4.2 Electron Penetration Range in g/m^2 as a Function of Electron Energy. *(Courtesy of Energy Sciences, Inc.)*

density material. Figure 4.3 illustrates depth–dose relations where R_{opt} (optimum depth) is the equal-entrance, equal-exit parameter; R_{50} is the depth where the exit dose is 50% of the maximum dose; R_{50e} is the depth where the exit dose is 50% of the entrance dose; and R_p is the depth where the tangent line at the inflection point of the decreasing curve intersects the depth axis [22]. Figure 4.4, in a simplified way, shows this dependence of EB penetration on electron energy in the mid- to high-energy area as a linear function using the equal-entrance, equal-exit criteria also in centimeters of water or unit density materials [23–26].

Figure 4.5 shows the simplified depiction of such data as a linear function using the equal-entrance, equal-exit criteria, in unit density and the depth of

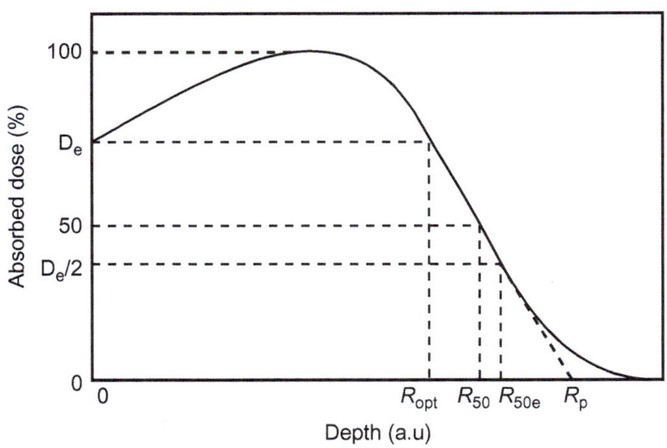

Figure 4.3 Useful Electron Ranges; R_{opt} is the "Equal-Entrance = "Equal-Exit" Depth.

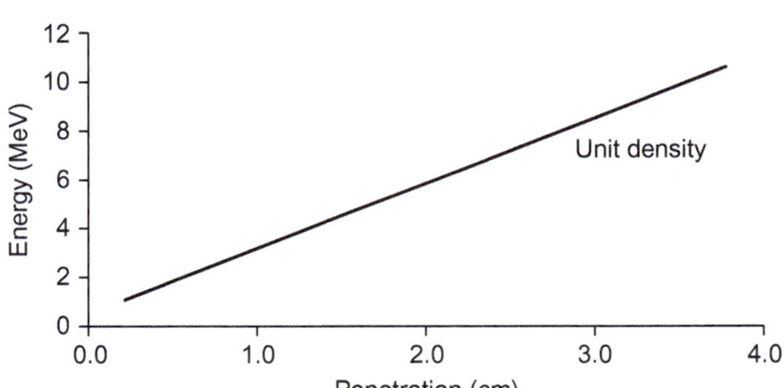

Figure 4.4 Electron Beam Penetration in Centimeters of Water.

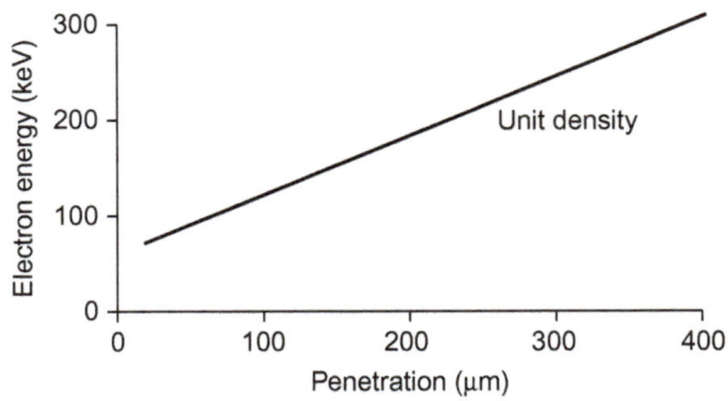

Figure 4.5 Low-Energy Electron Beam Penetration (Unit Density).

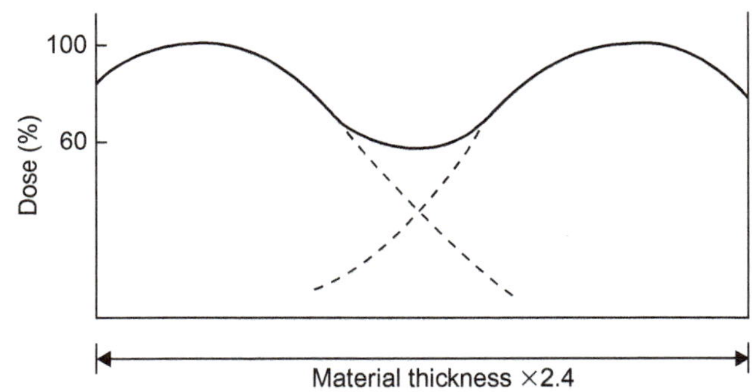

Figure 4.6 Illustration of Opposite-Sided EB Exposure.

penetration in micrometers (μm). In the low-energy EB area, coating coverage expressed in grams per square meter (g/m^2) is equal to thickness in μm for unit density, 1.0 g/cm^3, materials. Changes in formula density due to pigment loadings and the like require corrections to be made in proportion to the formula density; higher density formulations will have a proportionally lower thickness given the same area coverage.

Corrections for material density have to be made in order to assess the appropriate voltage for a given application. For example, fillers used in coating formulations and in wire and cable compounds will increase product density. Given an overlap of tail ends of the depth–dose penetration, opposite-sided EB exposure results in an effective 2.4 multiple of the EB penetration itself [27]. Thus, fairly large, low bulk density packages can be irradiated if the item is turned over during processing. Figure 4.6 illustrates this effect of opposite-sided EB treatment.

Experiments on cross-linking by irradiation with EB have revealed that in many cases the cross-link densities equivalent to those obtained by conventional methods of cross-linking require relatively high radiation doses.

Therefore, a great deal of experimental work has been done to determine methods that would enhance the efficiency of the EB radiation process. The currently known methods include addition of a sensitizer, postirradiation heat treatment, compression, high-temperature irradiation, addition of a plasticizer, the addition of a polyfunctional monomer, or the addition of certain fillers (see Section 4.2.2.3). These methods lead to increasing the number of polymer radicals in the amorphous region and increasing the probability of their recombination. A thorough study on the use of additives, which promote cross-linking by irradiation, has been performed [20, p. 593]. Such compounds are called radiation *cross-link promoters* or *prorads*.

4.2 Methods to Enhance EB Radiation Cross-linking by Prorads

The cross-link promoters (prorads) are prone to form more radicals and faster than polymers alone, thus, lowering radiation doses needed to achieve required properties. There are essentially two types of promoters of radiation cross-linking [20]. One group, *indirect cross-link promoters*, does not enter directly into the cross-linking reaction but merely enhance the formation of reactive species, such as free radicals, which then lead through secondary reactions to cross-linking. The other group, *direct cross-link promoters*, enters directly into the cross-linking reaction and become the actual connecting molecular links.

4.2.1 Indirect Cross-link Promoters

4.2.1.1 Halogenated Compounds

This group of compounds has been studied extensively [20, p. 593] [28–35], mainly in elastomers. The findings can be summarized as follows: For chlorinated aliphatic compounds, the sensitizing effect increases with decreasing number of carbon atoms in the molecule. The cross-link promoting effect increases on passing from iodo- to bromo- to chloro-substituted compounds and with increasing degree of halogenation. Halogen-containing aromatic compounds also represent good cross-linking promoters [20, p. 596].

4.2.1.2 Nitrous Oxide

Enhanced radiation cross-linking in polyethylene, polypropylene, and polyisobutylene [36] and in copolymers of ethylene and propylene [37] was found when nitrous oxide was incorporated into the polymer matrix. Mechanisms of this process have been proposed by several workers [20,38,39].

4.2.1.3 Sulfur Monochloride

Several investigators found moderate [40] to substantial [41] increase in the rate of cross-linking of polyethylene and polypropylene. It was speculated that the differences between the results of the two studies might be due to the differences in the way the promoter was distributed in the test material [20, p. 597].

4.2.1.4 Bases

The promotion of cross-linking of polypropylene [42] and ethylene–propylene copolymers [40] has been confirmed. However, other bases, such as amines, were not found effective; in some cases they acted in fact as retarders of radiation cross-linking [20, p. 600].

4.2.2 Direct Cross-link Promoters

4.2.2.1 Maleimides

Maleimides and dimaleimides are known to accelerate cross-linking of elastomers by organic peroxides at elevated temperatures [43], but they also have been found to sensitize radiation-induced cross-linking of polymers [20, p. 601, 603] [44]. Experiments involving radiation cross-linking of purified natural rubber and other elastomers have demonstrated that several maleimides, alkyl- and aryldimaleimides, significantly enhance the rate of cross-linking. Among the most effective were *N*-phenylmaleimide and *m*-phenylene dimaleimide, which in the amount of 5 wt% increased the $G(X)$ of purified natural rubber approximately by the factors of 23 and 15, respectively. The differences in the sensitizing action of the maleimides tested were at least attributed to their different solubility in the rubber. For the *m*-phenylene dimaleimide, the rate of cross-linking was found to be directly proportional to its concentration up to 10 wt% [20].

The mechanism of the cross-link promotion of maleimides is assumed to be based on copolymerization of the polymer *via* its unsaturations with the maleimide molecules initiated by radicals, and in particular, by allylic radicals produced during the radiolysis of the polymer [43].

When tested in other polymers, maleimides did not affect the rate of cross-linking in polydimethylsiloxane, polyisobutylene, and polyvinylchloride. In polyethylene, the addition of 5 wt% of *m*-phenylene dimaleimide increased the $G(X)$ from 1.8 to 7.2. In the polyvinylacetate, the effect was even more pronounced: the dose for gelation was reduced by about a factor of 50 [40]. Contrary to the cross-link enhancing effect found for *m*-phenylene dimaleimide, cross-linking was retarded in polyvinyl acetate by the addition of monomaleimides. When analyzing the mechanism of the reaction, it was concluded that monomaleimides are not expected to affect cross-linking in saturated polymers [45].

4.2.2.2 Thiols (Polymercaptans)

Studies of several investigators [46–50] have shown that polyfunctional thiol compounds are useful as cross-linking agents for the radiation curing of unsaturated elastomers and polymers used in graphic arts, electronics, and coating industries. Already small amounts of these compounds enhance the rate of cross-linking. They also promote cross-linking of polyisobutylene and its copolymers, which normally degrade on exposure to radiation [50].

The basis for cross-linking in the elastomers in these studies is the addition of thiols to olefinic double bonds. This reaction has been studied by a number of investigators [47–51]. It has been shown that the addition reaction proceeds by a radical mechanism. The rate of cross-linking of polybutadiene can be strongly promoted by relatively small amounts of polythiols [52]. For example, $G(X)$ values increased from 5 (pure polybutadiene) to 29 when 1 wt% of dimercaptodecane was added and to more than 49 on the further addition of *o*-dichlorobenzene (Table 4.2).

4.2.2.3 Acrylic and Allylic Compounds

As pointed out earlier, multifunctional monomers (prorads) essentially participate in a number of radical mechanisms, such as radical addition and grafting. Generally, they improve effectiveness of irradiation with EB by increasing cross-link density. Multifunctional monomers can be grouped according to their influence on cure kinetics and ultimate physical properties. Type I multifunctional monomers are highly reactive and increase both the rate and the state of cure. Monomers belonging to this group include acrylate, methacrylate, or maleimide functionality. They are very polar structures with limited solubility in most polymers.

Polyfunctional monomers, such as polyacrylic and polyallylic compounds, have been found to enhance the radiation cross-linking of polyvinyl chloride [53]. They were also found to accelerate cross-linking of elastomers; however, their effect is rather small [20, p. 604]. Radiation cross-linking of butadiene–styrene copolymers and natural rubber compounded with 50 phr (parts per hundred parts of rubber) of carbon black [54] in the presence of acrylate promoters exhibited greatly enhanced physical properties (modulus at 300% elongation and tensile strength). The best results were attained with tetramethyl diacrylate and ethylene

Table 4.2 Promotion of Cross-linking by Polythiols

Elastomer	Prorad	Prorad Amount (wt%)	$G(X)$ Physical	Note
Polybutadiene	–	–	5	Irradiation in air (2.5 kGy/s)
	Dodecanethiol	1.0	4.5	Irradiation in air (2.5 kGy/s)
	Dimercaptodecane	1.0	29	Irradiation in air (2.5 kGy/s)
	Dimercaptodecane +2 wt% *o*-dichlorobenzene	1.0	>49	
	Dipentene dimercaptan	1.0	18	
	α, α'-dimercapto-*p*-xylene	1.0	39	
	Trimethylolpropane trithioglycolate	1.0	>35	Irradiation in air (2.5 kGy/s)

dimethacrylate. Several polyfunctional allylic and acrylic compounds were evaluated in the radiation cross-linking of polyethylene, polypropylene, polyisobutylene, and ethylene–propylene copolymers. The rate of gelation of these polymers was increased considerably, but many of the radiation promoters homopolymerized on radiation exposure and, when compounded with elastomers, formed a rigid network structure within the more yielding elastomeric structure [40,55,56].

The efficiency of promoting the cross-linking of polymers depends on the solubility of the prorad in the polymer and its reactivity with the polymer radicals [57]. When the solubility of the prorad is sufficient, its efficiency of promoting cross-linking is proportional to its specific unsaturation (number of moles of double bonds per 100 g of the monomer [58]). The cross-link density ($1/M_c$) of the product is linearly proportional to the absorbed dose (D) and the initial prorad content (c_0):

$$1/M_c = (A + Kc_0)D \tag{4.3}$$

where A and K are constants; the constant K is dependent on the specific unsaturation.

Examples of different polyfunctional monomers effective in a variety of polymers are given in Table 4.3. Although triallyl cyanurate (TAC) and triallyl isocyanurate (TAIC) are very useful and effective cross-linking promoters, they are not versatile. Moreover, the use of TAC and TAIC is limited by their toxicity (TAIC is known as a mutagen) [57, p. 85].

It should be noted that the effectiveness of these prorads is influenced by other factors, such as temperature and presence of fillers (e.g., carbon black). In general, the cross-link density is increased by the addition of fillers. The cross-link enhancing activity of carbon black results from the induced physical and chemical cross-linking by carbon black. The physical cross-linking is based on the entanglements of the polymeric chains and the surface porosity of the carbon black. The chemical cross-linking is formed by radical reactions between the carbon black and the polymer molecules. Other fillers enhancing cross-linking are zinc oxide, magnesium oxide, and silica fillers [57, p. 87]. A sufficiently large amount of added filler interferes with crystallization of the polymer. Since radiation cross-linking depends on the volume fraction of the amorphous region, a small amount of filler acts as a nucleacting agent, while a large amount enhances the radiation cross-linking [59].

Other known methods of increasing radiation cross-linking efficiency are [57, p. 72] postirradiation heat treatment and high-temperature irradiation.

Table 4.3 Examples of Effective Polyfunctional Prorads in Various Polymers

Polymers	**Effective Polyfunctional Prorad(s)**
Elastomers	
Styrene–butadiene copolymer (SBR)	Tetramethylol methane tetracrylate, tetramethylene diacrylate, ethylene dimethylacrylate
Chlorobutyl (CIIR)	Trimethylolpropane trimethacrylate
Polyisoprene rubber (IR)	Diethyleneglycol dimethacrylate
Polychloroprene (CR)	Polyethyleneglycol dimethacrylate
Ethylene–propylene rubber (EPM)	Triethyleneglycol dimethacrylate
Ethylene–propylene–diene rubber (EPDM)	Ethyleneglycol dimethacrylate
Butadiene–acrylonitrile rubber (NBR)	Diethyleneglycol dimethacrylate, trimethylolpropane trimethacrylate
Fluorocarbon elastomer (FKM)	Trimethylolpropane trimethacrylate, trimethylolpropane triacrylate
Natural rubber (NR)	Diethyleneglycol dimethacrylate
Conventional plastics	
Polyethylene (PE)	Triallyl cyanurate, allyl methacrylate
Polypropylene (PP)	Trimethylolpropane triacrylate
Polyvinyl chloride (PVC)	Polyethyleneglycol dimethacrylate, triallylcyanurate
Polyvinylidene fluoride (PVDF)	Triallyl trimellitate
Ethylene vinyl–acetate copolymer (EVA)	Trimethylolpropane triacrylate, triallyl isocyanurate
Engineering plastics	
Polyamide (PA) 6, 66, 12	Triallyl isocyanurate
Polyamide (PA) 610	Triallyl isocyanurate
Polybutylene terephthalate (PBT)	Triallyl isocyanurate
Biodegradable plastics	
Polycaprolactone (PCL)	Trimethallyl isocyanurate
Polybutylene succinate (PBS)	Trimethallyl isocyanurate

Postirradiation heat treatment (annealing) increases the number of polymer radicals in the amorphous region to facilitate the migration of trapped radicals in the crystalline regions to the amorphous regions. An example of this process is the cross-linking of polypropylene, polyvinylidene fluoride, and polyvinyl acetate.

High-temperature irradiation increases the recombination of polymer radicals. There is a specific critical temperature for each polymer below which migration and recombination of radicals is suppressed. Typically it is always above the glass transition temperature. Examples of the use of high-temperature are PVC, PVA, and FEP. Melted state irradiation is favorable for PTFE. Radiation cross-linking of PTFE occurs under inert atmosphere just above its crystalline melting temperature.

4.3 Retarders of Radiation Cross-linking

Retarders of cross-linking often referred to as *antirads* were first found to retard cross-linking but also to protect polymers against radiation damage [20, p. 606]. A considerable reduction of cross-linking was found for a number of aromatic amines, quinones, aromatic hydroxyl sulfur compounds, and aromatic nitrogen compounds. The degree of protection offered by these compounds increases with their concentration but reaches a limiting value at a concentration of a few weight percent [20, p. 606].

The above-mentioned radical acceptors are not the only effective retarders of cross-linking but are also protective agents against radiation. Studies performed with benzene and nitrobenzene in natural rubber [35] and SBR with *N*-phenyl-β-naphthylamine [60] demonstrated radiation protection of the respective compounds.

Many of the studied radical-accepting compounds were found to reduce the yield of scissions [61,62]. The antirads tested prove effective not only in the presence of air but also in its absence (Table 4.4). The subject of protective agents is discussed in more detail in Chapter 7.

Different polymeric materials respond to irradiation by EB in different ways. A large number of them will be modified by the formation of a cross-linked network; by changing their surface properties or structure, some will be degraded. Another field of ionizing radiation applicable to polymeric systems is grafting. EB irradiation can also be used for polymerization and/or cross-linking of oligomers and monomers, i.e., in conversion of liquids to solids. These subjects will be discussed to a greater extent in Chapter 5.

Table 4.4 Effect of Antirads on the Yield of Cross-links and Scissions in Natural Rubber

Natural Rubber Compound (50 phr EPC Carbon Black) + 5 phr Antirad	*G*(*X*)		*G*(*S*)[a]	
	Vacuum	Air	Nitrogen	Air
Control (1 phr phenyl-2-naphthylamine)	1.9	0.29	2.7	13
N-*N'*-Dicyclohexyl-*p*-phenylenediamine	–	–	1.5	3
N-Cyclohexyl-*N'*-phenyl-*p*-phenylene diamine	1.3	0.33	1.2	1.4
6-Phenyl-2,2,4-trimethyl-1,2-dihydroquinoline	0.83	0.19	1.9	4.2
N-*N'*-Dioctyl-*p*-phenylene diamine	0.87	0.12	1.5	5.0
2-Naphthylamine	0.87	0.30	1.6	5.6
p-Quinone	–	–	2.8	7.8
Phenylhydroquinone	1.1	0.46	2.2	5.4
1,4-Naphthoquinone	1.1	0.48	2.0	5.6
2-Naphthol	1.1	0.24	1.3	4.1
N-*N'*-Diphenyl-*p*-phenylenediamine (35%) and phenyl-1-naphthylamine (65%)	0.97	0.27	1.4	3.7

[a] *Scission yields are determined by stress relaxation measurements during irradiation of specimens. Permanent and temporary chain scissions are measured by this technique in contrast to gel measurements, which provide data only on permanent scissions.*

References

[1] Kase KR, Nelson WR. Concepts of radiation chemistry. New York, NY: Pergamon Press; 1978.

[2] Davidson RS. Exploring the science, technology and applications of U.V. and E.B. curing. London: SITA Technology; 1999. p. 123.

[3] Lowe C. In: Oldring PKT, editor. Chemistry and technology of U.V. and E.B. formulations for coatings, inks and paintings, vol. 4, [chapter 1]. London: SITA Press; 1991.

[4] Schafer O, Allan M, Haselbach E, Davidson RS. Photochem Photobiol 1989;50:717.
[5] Clegg DW, Collier AA, editors. Irradiation effects on polymers. London: Elsevier; 1991.
[6] Singh A, Silverman J, editors. Radiation processing of polymers. Munich: Hanser Publishers; 1992.
[7] Mehnert R. Radiation induced polymerization. In Ullmann's encyclopedia of industrial chemistry, vol. A22. Weinheim: VCH; 1993. p. 471.
[8] Garratt P. Strahlenhärtung. Hannover: Curt Vincentz Verlag; 1996 [in German].
[9] Tabata T, Ito R. An algorithm for the energy deposition of fast electrons. Nucl Sci Eng 1976;53:226.
[10] Frankewich EL. Usp Khim 1966;35:1161.
[11] Hirsch J, Martin E. Solid State Commun 1969;7:279–783.
[12] Dole M, Böhm GGA, Waterman DC. *Polym J* 1969;vol.1(Suppl. 5):93.
[13] Mehnert R, Pincus A, Janorsky I, Stowe R, Berejka A. UV&EB technology and equipment, vol. 1. London and Chichester: SITA Technology, Ltd./John Wiley & Sons; 1998.
[14] Buijsen P. Electron beam induced cationic polymerization with onium salts (Thesis, Delft University of Technology). Delft University Press; 1996.
[15] von Raven A, Heusinger H. J Polym Sci 1974;12:2235.
[16] Kaufmann R, Heusinger H. Makromol Chem 1976;177:871.
[17] Katzer H, Heusinger H. Makromol Chem 1973;163:195.
[18] Zott H, Heusinger H. Makromolekules 1975;8:182.
[19] Kuzminskii AS, Bolshakova SI. Symp. Rad. Chem. 3rd. Tihany, Hungary; 1971.
[20] Böhm GGA, Tveekrem JO. The radiation chemistry of elastomers and its industrial applications (review). Rubber Chem Technol 1982;55 (3):592.
[21] Lapin S. Electron beam-activated cationic curing. Paper presented at RadTech 2012. Chicago, IL; 2012.
[22] ISO/ASTM standard 51649 standard practice for dosimetry in an electron beam facility for radiation processing at energies between 300 keV and 25 MeV. West Conshohocken, PA: ASTM International.
[23] Cleland MR. In: Singh A, Silverman J, editors. Radiation processing of polymers [Chapter 3]. Munich: Carl Hanser Verlag; 1992. p. 23.
[24] Cleland MR. J Ind Irradiation Technol 1983;1(3):191.
[25] Cleland MR. Industrial applications of electron accelerators. Zeegse, The Netherlands: CERN Accelerator School; 2005.
[26] Cleland MR, Galloway RA, Berejka AJ. Energy dependence of electron beam penetration, area throughput rates and electron energy utilization in the low energy region. Nucl Instr Meth B 2007;261:94.

[27] Bly JH. Electron beam processing. Yardley, PA: International Information Associates; 1988. p. 28.
[28] Chapiro AJ. Chim Phys 1950;47:747–64.
[29] Dogadkin BA, Miadenov A, Tutorskii IA. Vysokomol Soedin 1960;2:259.
[30] Jankowski B, Kroh J. J Appl Polym Sci 1969;13:1795.
[31] Jankowski B, Kroh J. J Appl Polym Sci 1965;9:1363.
[32] Kozlov VT, Kaplanov MYA, Tarasova ZN, Dogadkin BA. Vysokomol Soedin 1968;A10(5):987.
[33] Kozlov VT, Klauzen NA, Tarasova ZN. Vysokomol Soedin 1968; A10(7):1949.
[34] Nikolskii VG, Tochin VA, Buben NYa. Elementary processes of high energy. Izd Nauka 1965; [in Russian].
[35] Turner D. J Polym Sci 1958;27:503.
[36] Okada Y. Adv Chem Ser 1967;66:44.
[37] Karpov VI. Vysokomol Soedin 1965;7:1319.
[38] Scholes G, Sinic M. Nature 1964;202:895.
[39] Lyons BJ, Dole M. J Phys Chem 1964;68:526.
[40] Geymer DO, Wagner CD. Polym Prep Am Chem Soc Div Polym Chem 1968;9:235.
[41] Kaurkova GK, Kachan AA, Chervyntsova LL. Vysokomol Soedin 1965;7:183, *J Polym Sci Part C* 1967;16:3041.
[42] Geymer D. Macrom Chem 1967;100:186.
[43] Miller SM, Roberts R, Vale LR. J Polym Sci 1962;58:737.
[44] Miller SM, Spindler MW, Vale LR. Proc IAEA Conf Appl Large Rad Sources Ind Salzburg 1963;1:329.
[45] Pearson DS, Shurpik A. U.S. patent 3646502 to Firestone Tire and Rubber; 1974.
[46] Zapp RL, Oswald AA. Paper #55. Meeting of the Rubber Division of American Chemical Society. Cleveland.
[47] Griesbaum K. Angew Chem Ind Ed Engl 1970;9(4):273.
[48] Morgan CR, Magnotta F, Kelley AD. Thiol/ene photocurable polymers. J Polym Sci Polym Chem 1977;15:627.
[49] Morgan CR, Kelley AD. J Polym Sci Polym Lett 1978;16:75.
[50] Pierson RM, Gibbs WE, Meyer GE, Naples FJ, Saltman WM, Schrock RW, Tewksbury LB, Trick GS. Rubber Plast Age 1957;38:pp. 592, 708 and 721.
[51] Walling C, Helmreich W. J Am Chem Soc 1978;81:1144.
[52] Böhm GGA. In: Dole M, editor. The radiation chemistry of macromolecules, vol. II, [chapter 12]. New York, NY: Academic Press; 1972.
[53] Miller AA. Ind Eng Chem 1959;51:1271.
[54] Smith WV, Simpson VG. U.S. patent 3084115 to U.S. Rubber Co; 1963.

[55] Lyons BJ. Nature 1960;185:604.
[56] Odian G, Bernstein BS. Nucleonics 1963;21:80.
[57] Makuuchi K, Cheng S. Radiation processing of polymer materials and its industrial applications. Hoboken, NJ: John Wiley & Sons; 2012. p. 84.
[58] Xu Y, Fu Y, Yoshii F, Makuuchi K. Rad Phys Chem 1998;53:669.
[59] Mateev M, Nikolova M. Polym Degrad Stabil 1990;30:205.
[60] Blanchford J, Robertson PF. J Polym Sci 1965;3:1289, 1303, 1313 and 1325.
[61] Bauman RG, Born JW. J Appl Polym Sci 1959;1:351.
[62] Bauman RG. J Appl Polym Sci 1959;2:328.

Recommended Further Reading

Makuuchi K, Cheng S. Radiation processing of polymer materials and its industrial applications. Hoboken, NJ: John Wiley & Sons; 2012.
Industrial Radiation Processing with Electron Beams and X-Rays. International Atomic Energy Agency. Vienna, Austria; 2011.
Drobny JG. Radiation technology for polymers. Boca Raton, FL: CRC Press; 2010.
Clough RL, Shalaby SW. Irradiation of polymers: fundamentals and technological applications [ACS Symposium Series 620]. Washington, DC: American Chemical Society; 1996.
Garratt PG. Strahlenhärtung. Hannover: Curt R. Vincentz Verlag; 1996 [in German].
Mehnert R. Radiation chemistry: radiation induced polymerization in Ullmann's encyclopedia of industrial chemistry, vol. A22. Weinheim: VCH; 1993.
Singh A, Silverman J, editors. Radiation processing of polymers. Munich: Carl Hanser, Verlag; 1992.
Clough R. In: Mark HF, Kroschwitz JI, editors. Encyclopedia of polymer science and engineering, vol. 13. New York, NY: John Wiley & Sons; 1988.
Bradley R. Radiation technology handbook. New York, NY: Marcel Dekker; 1984.
Charlesby A. Atomic radiation and polymers. New York, NY: Pergamon Press; 1960.

5 Electron Beam Processing of Commercial Polymers, Monomers, and Oligomers

OUTLINE

This chapter covers the subject of processing of plastics, elastomers, inks, adhesives, and coatings using EB (electron beam) equipment. Plastics are by far the largest group of polymeric materials being processed by EB irradiation. Cross-linking of polyolefins, PVC, polyesters, polyurethanes, fluoropolymers, and fiber-reinforced composites is a common practice. In addition, liquid systems containing monomers and oligomers are frequently processed by EB radiation.

5.1 Processing of Thermoplastics

5.1.1 Processing of Polyolefins

Irradiation of the family of polyolefins (PEs) represents an important segment of radiation processing. Polyolefins can be irradiated in many forms, such as pellets and powders, films, extruded, and molded parts or as wire and cable insulation.

Pellets and powders as well as small parts are handled simply by conveyor belts; cart systems or air transport tubes are the most common transportation systems for continuous EB irradiation [1]. Wire and cable as well as tubular products require special handling. Frequently, irradiation from one side is not sufficient since it results in nonuniform dose distribution. Therefore, the

Drobny: Ionizing Radiation and Polymers.
DOI: http://dx.doi.org/10.1016/B978-1-4557-7881-2.00005-5

product has to be twisted so that its entire circumference is presented to the source of irradiation. Multiple passing is another way to alleviate this problem. Details are given in Chapter 6 in sections on wire and cable (Section 6.1) and heat shrinkable products (Section 6.4). Thick sheets and extruded profiles have to be often irradiated from both sides, passing under the beam several times. Large and thick parts require spinning along some predetermined axis or angle to assure uniform dose distribution [1].

5.1.1.1 Polyethylene

The effects of ionizing radiation on PE in all its forms can be summarized as follows [1, p. 105]:

- The evolution of hydrogen
- The formation of carbon–carbon cross-links
- An increase in unsaturation to an equilibrium level
- A reduction in crystallinity
- The formation of color bodies in the resin
- Surface oxidation during irradiation in air.

The formation of carbon–carbon cross-links is by far the most important effect and is the basis of the applications in wire and cable industry and for heat-shrinkable products. The factors affecting the changes of PE by irradiation are the molecular weight distribution, branching, the degree of unsaturation, and morphology [2].

Low-density polyethylene (LDPE), produced by a high-pressure polymerization, contains long branches attached to the main chain. Linear low-density polyethylene (LLDPE) has a rather regular succession of short-chain branches. Both these types have a small amount of unsaturation, but high-density linear polyethylene (HDPE) contains one terminal vinyl group per molecule. At low-radiation doses, this vinyl group has the effect of increasing the molecular weight of the HDPE by the chain formation of Y-links between the vinyl and the secondary alkyl radicals produced by the radiation [3].

At ambient temperatures, PE is always in the semicrystalline form. The amorphous region has a density of 800 kg/m^3 (50 lb/ft^3), while the density of the crystalline region is 25% greater. The range of commercially available PEs is from 920 to 960 kg/m^3 (58 to 60 50 lb/ft^3), but this relatively small difference in density corresponds to a considerably greater difference in the amorphous fraction; its weight fraction in LDPE is 40% and about 20% in HDPE [3, p. 17]. The cross-linking takes place mainly in the amorphous region and the interface between the two phases. The cores of the crystalline

Table 5.1 Yields of Cross-linking for a Range of PEs

Polyethylene Resin	Density	$G(X)$
LDPE	0.920	1.09
LDPE	0.935	0.8
HDPE	0.962	1.0
HDPE	0.950	0.70
HDPE	0.945	0.50
LLDPE	0.937	1.0
LLDPE	0.924	0.96
LLDPE	0.919	0.99

regions sustain radiation-induced transvinylene formation in proportion to their weight fraction, but they are scarcely involved in gel formation [3, p. 17]. Radiation cross-linking effects and the degree of cross-linking within crystalline regions are reported in Refs [4–8]. Cross-link yields, i.e., $G(X)$ of a wide range of PEs, are given in Table 5.1.

Cross-linking of PE by irradiation takes place normally at temperatures below 70°C (158°C), which is below its alpha-transition temperature and much lower than the crystalline melting point. Irradiation does not have any significant effect on sizes of amorphous and crystalline fractions as measured by the heats of fusion. However, when the cross-linked polymer is heated above the crystalline melting point, the crystallinity is significantly reduced because the cross-links interfere with the reformation of the supermolecular fraction during cooling from the melt. Subsequent cycles of melting and cooling do not produce additional changes in crystallinity [3, p. 17] [8,9].

The core of the crystalline region of irradiated PE contains residual-free radicals. These diffuse slowly to the interface with the amorphous region, where, in the presence of dissolved oxygen, whose equilibrium concentration is maintained by diffusion, they initiate an autooxidative chain of degradation [10]. Postirradiation annealing in an inert atmosphere at a temperature above the alpha-transition temperature (85°C or 185°F) leads to a rapid mutual reaction of the free radicals and eliminates the problem [3, p. 18].

The higher crystalline fraction of the radiation cross-linked PE, even after a melt-freeze cycle, has great technological merit for heat-shrink packaging and electrical connector products [3, p. 19, p. 20].

Compared with chemical cross-linking of PE, radiation curing produces a different product in many respects. Chemical cross-linking is done at temperatures near 125°C (257°F), where the polymer is in the molten state. Consequently, the cross-link density in the chemically cross-linked PE is almost uniformly distributed, while there are relatively few cross-links in the crystalline fraction of the radiation cross-linked PE. The crystalline fraction of the radiation-processed PE is greater than that in the chemically cured product [3, p. 20].

Radiation cross-linking of PE requires considerably less overall energy, requires less space and is faster, more efficient, and environmentally more acceptable [11]. Chemically cross-linked PE contains chemicals that are by-products of the curing system. These often have adverse effects on the dielectric properties and in some cases are simply not acceptable [12].

Cross-linking of PE enhances its resistance to plastic flow and resistance to wear. Wear resistance of HDPE is enhanced by irradiation in vacuum and reduced by irradiation in air [13]. This finding is used to increase the wear resistance of ultrahigh molecular weight polyethylene (UHMWPE) used for orthopedic implants [14]. In general, the wear resistance of PE increases with the radiation dose and cross-link density and is favorable for PE-based wire and cable insulation [15].

Another benefit of the introduction of cross-links into the polymer is an improvement of its chemical resistance. Solubility and swelling decrease with increasing gel content of the polymer as does its environmental stress-cracking resistance [16]. The increase of environmental stress-cracking resistance is particularly beneficial for pipes under constant pressure.

Dielectric constant values of LDPE are essentially independent of the degree of cross-linking; however, the loss factor increases, very likely due to formation of some polar groups during irradiation in the presence of air [17].

Melt flow onset temperature (a temperature at which a polymer starts to melt and flow) increases with increasing gel content up to 10%, then levels off. High values of the hot melt onset temperature of the cross-linked PE indicate very favorable properties at high temperatures. The creep of PE at high temperatures is greatly reduced with a high degree of cross-linking [18].

The disadvantage of EB cross-linking is a more or less nonuniform dose distribution. This can happen particularly in thicker objects due to intrinsic dose–depth profiles of EBs. Another problem can be a nonuniformity of rotation of cylindrical objects as they traverse a scanned EB. However, the mechanical properties often depend on the mean cross-link density [19].

5.1.1.2 Polypropylene

Polypropylene (PP) is a stereospecific polymer prepared by polymerization using an organometallic catalyst system. Commercial PPs have up to 95% isotactic content, which means that pendant methyl groups are almost all on the same side of the chain. Isotactic PP is a semicrystalline polymer, although the atactic form of PP is purely amorphous.

Table 5.2 Effect of Absorbed Dose on Mechanical Properties of Commercial PP

Property	Absorbed Dose (kGy)					
	0	100	280	800	1200	1600
Tensile strength (MPa)	37.5	35.1	30	17.1	18	16.5
Modulus of elasticity (MPa)	1.4	1.3	1.3	1.2	1.2	–
Elongation at break (%)	900	200	90	50	40	20

Table 5.3 *G*-Factors for Atactic and Isotactic PPs

Polymer	$G(X)$	$G(S)$	$G(S)/G(X)$
Atactic PP	0.27	0.22	0.8
Isotactic PP	0.16	0.24	1.5

When PP is exposed to ionizing radiation, free radicals are formed and these cause chemical changes. Since commercial PP is highly crystalline, these radicals are relatively immobile and consequently may not be available for reaction for long periods of time [1, p. 114].

As with other polyolefins, upon irradiation, the free radicals are formed along with evolution of hydrogen gas. If the radical is formed on the pendant methyl, the resulting reaction is cross-linking. However, if the radical is formed in the main chain, the chain end may react with hydrogen thus causing an irreversible scission. Although the processes of chain scission and cross-linking occur simultaneously, and even though the net effect is cross-linking, the overall effect is the loss of mechanical strength [1, p. 115]. This is particularly true for irradiation with doses above 500 kGy (Table 5.2). *G*-factors for cross-linking and scission for the two types of PPs are given in Table 5.3.

Irradiation also shows an effect on the level of crystallinity and melting point. For example, after a dose of 6000 kGy, crystallinity was found to be 73% of the original value and the melting point changed from 160°C to 105°C [20].

5.1.2 Polystyrene

Polystyrene is a clear, amorphous polymer with a high stiffness and good dielectric properties. It is easily cross-linked by ionizing radiation. Often small amounts of divinyl benzene (DB) may enhance the degree of cross-linking [1, p. 132].

Moderate doses, between 10 and 200 kGy (1.0 and 20 Mrad) on partially or completely formed articles and polymer pellets, substantially reduce the content of residual monomer [21]. The mechanical properties of polystyrene are changed only at high radiation doses, which is characteristic of low cross-link yield and glassy morphology [22]. The hardness, tensile strength, and shear strength are all within 75% of the original values up to 10^5 kGy [23]. The glass transition temperature is reported to increase to 150°C and the crystalline melting point increases to 150°C after the irradiation of the crystalline isotactic polystyrene to 4×10^4 kGy [24]. Typical crosslink yields, $G(X)$, are in the range 0.019–0.051, depending on the method of determination. Polystyrene is fairly resistant to the effects of ionizing radiation due to the presence of aromatic groups in its molecule.

5.1.3 Polyvinyl Chloride and Polyvinylidene Chloride

PVC is a slightly crystalline polymer, because of its branched structure. A large amount of PVC is used for insulation in the wire and cable industry. Radiation cross-linking of PVC requires the addition of a multifunctional prorad. In the absence of air, cross-linking predominates over chain scission and $G(X) = 0.33$. Addition of plasticizers or raising temperature above the glass transition, increases cross-link yield [25]. Irradiated PVC materials usually show increased solvent, heat, and flow resistance [26]. Radiation curable PVC and its copolymers are suitable for grafting involving multifunctional monomers [27].

The most effective prorads for PVC are acrylic and allylic esters, such as TAC and trimethylolpropane trimethacrylate (TMPTMA) and trimethylolpropane triacrylate (TMPTA). The triacrylate is more reactive than the trimethacrylate, but it is more toxic and for that reason is used only seldom. The amounts of these additives are 1–5% of the formulation weight [1, p. 134].

Degradation of PVC during irradiation with γ-rays or EB occurs via the loss of hydrogen chloride gas (dehydrochlorination), which is accompanied by severe color change to a dark brown material. The color is caused by the formation of highly conjugated double bonds. The degradation is more pronounced in the presence of air and continues after the irradiation has stopped [28]. The dehydrochlorination process is dependent on temperature and increases with increasing temperatures as does the postirradiation effect.

The physical properties of PVC film show an increase in elongation at low doses (100 kGy) and then a dramatic fall off in elongation at 300 kGy. Above doses of 300 kGy, the material becomes brittle [29].

Polyvinylidene chloride (PVDC) is known for its very good gas barrier properties. PVDC not only discolors upon irradiation but also degrades to the point that it is not useful. Therefore, it should not be used as an inner layer of irradiated packaging films.

5.1.4 Polymethacrylates and Polyacrylates

Poly(methyl methacrylate) (PMMA) degrades under irradiation and becomes more soluble due to main chain scission [30]. The degradation can be greatly reduced by the addition of 10% of various additives, such as aniline, thiourea, or benzoquinone [31]. PMMA is an example of a nongelling polymer; it does not form a three-dimensional network structure under irradiation [22, p. 442]. On the other hand, poly(alkyl acrylates) undergo radiation cross-linking. Tables 5.4 and 5.5 give *G*-factors for a series of poly(methacrylates) and poly(alkyl acrylates), respectively.

5.1.5 Fluoroplastics

There is a relationship between the effects of ionizing radiation on a fluoropolymer and the amount of hydrogen atoms in the monomeric unit. The trend for cross-linking can be approximately expressed as follows:

$$\text{PVF} > \text{PVDF} > \text{ETFE} > \text{FEP} > \text{PFA} > \text{PTFE}$$

Table 5.4 *G*-Factors for a Series of Poly(alkyl methacrylates)

Polymer	$G(X)$	$G(S)$	$G(S)/G(X)$
Poly(methyl methacrylate)	–	1.63 (*in vacuo*)	–
Poly(methyl methacrylate)	–	0.77 (in air)	–
Poly(phenyl methacrylate)	–	0.44	–
Poly(benzyl methacrylate)	–	0.14	–

Table 5.5 *G*-Factors for a Series of Poly(alkyl acrylates)

Polymer	$G(X)$	$G(S)$	$G(S)/G(X)$
Poly(methyl acrylate)	0.5	–	0.07
Poly(ethyl acrylate)	0.07	0.07	0.23
Poly(*n*-butyl acrylate)	0.21	–	0.14
Poly(isobutyl acrylate)	–	–	0.07

and for degradation:

$$PTFE > PFA \sim FEP > ETFE > PVDF > PVF$$

In general, the higher the hydrogen content, the higher the tendency of the fluoropolymer to cross-link. The presence of hydrogen leads to dehydrohalogenation (loss of hydrofluoride, HF) upon irradiation. The use of prorads is of advantage because high levels of cross-linking can be achieved without compromising the thermal stability of the polymer [32].

5.1.5.1 Polytetrafluoroethylene

Under normal conditions, polytetrafluoroethylene (PTFE) undergoes chain scission. However, there is evidence that irradiation of PTFE at temperatures above its melting point (e.g., at 603–613 K, or 626–644°F) in vacuum results in a significant improvement in tensile strength and elongation at 473 K (392°F) and in the tensile modulus at room temperature [33–35]. These findings strongly contrast with the greatly reduced properties after irradiation at lower temperatures than that. This clearly indicates cross-linking in the molten state, similar to effects of irradiation of PE. At temperatures above 623 K (662°F), thermal depolymerization is increasingly accelerated by irradiation and predominates over cross-linking at yet higher temperatures [36]. The propensity of PTFE to degrade under ionizing radiation with a substantial reduction of molecular weight is exploited commercially by converting PTFE scrap into a low molecular weight product that is then used in the form of very fine powders ("micropowders") as additive to inks and lubricants (for details see Section 6.5.5).

When PTFE is degraded by ionizing radiation under vacuum, the amount of toxic compounds, such as HF and CO, decreases. The threshold value of absorbed dose for vacuum degradation of PTFE was found to be about 50–60 kGy [37]. The dependence of molecular weights and high-temperature viscosity on dose is given in Table 5.6 [38].

Table 5.6 Molecular Weights and High-Temperature Viscosity of Vacuum Irradiated PTFE

Dose (kGy)	M_n ($\times 10^6$)	Viscosity at 380°C (Pa s)
0	>10	3.2×10^{10}
150	2.5	2.8×10^{8}
750	2.1	1.4×10^{7}
750[a]	0.9	8×10^{5}

[a] *Air sintered material; other materials were vacuum sintered.*

5.1.5.2 FEP and PFA

FEP, copolymer of tetrafluoroethylene (TFE) and hexafluoropropylene (HFP), and PFA, copolymer of TFE and perfluoropropylvinyl ether, have physical and chemical properties similar to those of PTFE but differ from it in that they can be processed by standard melt-processing techniques.

When FEP is exposed to ionizing radiation at ambient temperature, it degrades like PTFE with resulting reduction of physical properties. However, if the temperature of the polymer is raised before irradiation above its glass transition temperature, cross-linking predominates and manifests itself by the increase in viscosity. With doses greater than 26 kGy (2.6 Mrad), ultimate elongation and resistance to deformation under load at elevated temperatures are improved, and the yield stress is increased. However, the improvements are offset by some loss in toughness [39]. The radiation tolerance of FEP is up to 10 times greater than that of PTFE.

PFA exhibits a considerably higher tolerance to ionizing radiation than PTFE, but it undergoes predominantly chain scission accompanied by a steady reduction of its mechanical properties [40].

5.1.5.3 Other Fluoroplastics

Polychlorotrifluoroethylene (PCTFE) according to one source degrades upon ionizing irradiation in a similar fashion to PTFE at ambient and elevated temperatures. Unlike PTFE, when irradiated above its crystalline melting point, it still exhibits chain scission [41]. However, the resistance of PCTFE to ionizing radiation is reported to be better than that of other fluoropolymers [41].

Copolymer of ethylene and tetrafluoroethylene (ETFE): This alternating copolymer can be cross-linked by irradiation [42]. Further improvement is achieved with the use of prorads, such as TAC or TAIC in amounts up to 10%. Irradiation at room temperature followed by heat treatment at 162°C in nitrogen for 20 min indicates some cross-linking [42]. Tensile properties of irradiated ETFE measured at 200°C are listed in Table 5.7. Irradiated ETFE insulation of wires exhibits improved heat resistance and resistance to cut-through. ETFE wire insulation is widely used in military and commercial aircraft.

Copolymer of ethylene and trichlorofluoroethylene (ECTFE) behaves upon irradiation like ETFE including improvement of cross-linking efficiency with prorads. Both the tensile strength and elongation increase up to a point but then start to fall. The room temperature properties show a maintenance of the tensile strength even up to 700 kGy but accompanied by a steady decrease in elongation [29].

Table 5.7 Tensile Properties of Irradiated ETFE, Measured at 200°C

Dose (kGy)	Temperature of Irradiation (°C)	Tensile Yield Strength (MPa)	Tensile Strength (MPa)	Elongation at Break (%)
0	–	2.4	2.4	12
7	RT[a]	3.7	5.8	545
7	150–198	3.7	5.6	421
10	220–245	3.2	4.8	340

[a]*Irradiation followed by treatment at about 160°C for 20 min.*

Table 5.8 Yields of Cross-linking and Scission for PVDF After Irradiation

Polymer	$G(X)$	$G(S)$	$G(S)/G(X)$	Remarks
PVDF (homopolymer)	1.0	0.3	0.30	
PVDF (homopolymer)	0.78	0.37	0.47	Solef 1010[a]
PVDF (homopolymer)	0.78	0.8	1.03	KF 1000[a]
PVDF (homopolymer)	0.75	0.77	1.03	KF 1100[a]
PVDF (homopolymer)	0.90	0.85	0.95	Kynar 200[a]
PVDF (homopolymer)	0.70	0.57	0.81	Kynar 450[a]

[a]*Irradiation at 61°C.*

Polyvinylidené fluoride (PVDF) cross-links upon irradiation, particularly with the use of prorads, such as TAC, TIAC, diallyl itaconate, and ethylene bis-maleimide and others [41]. At relatively low doses (less than 300 kGy), there is virtually no difference in the room temperature tensile properties when PVDF is irradiated with EB. For higher doses, such as above 300 kGy, there is an increase in the Young's modulus and decrease in both tensile strength and elongation at break [43]. Yields of cross-linking and scission for several PVDF homopolymers after irradiation are given in Table 5.8. The crystallinity of PVDF films has been shown to increase after irradiation with EB followed by aging at ambient temperature and uniaxial orientation [44].

High-energy irradiation of the piezoelectric β-crystalline form of PVDF reduces its piezoelectric potential and leads to improvement of thermal stability of the β-form and a slower piezoelectric decay [45].

Polyvinylfluoride (PVF) undergoes predominantly cross-linking when exposed to high-energy radiation [46], with a $G(X)$ of 3.4–5.7, $G(S)$ of

0.95–1.6, and *G(S)*/*G(X)* of 0.28. The tensile strength of PVF almost doubles upon irradiation of 10 kGy indicating the predominance of cross-linking [40].

Polytrifluoroethylene (PF_3E) undergoes both cross-linking and chain scission with the former dominating. The *G(X)* and *G(S)* values are 1.1 and 0.4, respectively [47].

5.1.6 Engineering Plastics

Engineering thermoplastics are a group of plastic materials that exhibit superior mechanical and thermal properties under a wide range of conditions over and above more commonly used commodity plastics. The term usually refers to thermoplastics rather than to thermosetting ones. Depending on performance, we can divide them to two categories: (1) regular service engineering plastics and (2) high-performance engineering plastics.

5.1.6.1 Regular Service Engineering Plastics

Polyamides cross-link when irradiated by EB or γ-rays. Both cross-linking and scission occur; the yields have been shown to be independent of the absorbed dose [48] but have found to be dependent on the number of hydrogen atoms or methylene groups in the amine residue [49].

In general, copolymers cross-link more readily than polyamide PA 66. Mechanical properties of polyamides are modified by irradiation as seen by reduced tensile strength (50% loss when irradiated in air, 16% under vacuum) [22, p. 443]. Aromatic polyamides retain strength better than aliphatic polyamides [50]. The major benefit of cross-linking of polyamides is their increased heat resistance.

Aliphatic polyesters tend to cross-link upon irradiation. Poly(ethylene terephthalate) (PET) cross-links with a low-efficiency; *G(X)* factors are between 0.035 and 0.14 and *G(S)* factors between 0.07 and 0,17 for dose rates of 90–130 kGy/min. However, it can sustain desirable physical properties up to 0.5 MGy [22, p. 443]. Beyond that the tensile strength and elongation values decrease and beyond 1 MGy the mechanical properties are destroyed [22, p. 443]. Poly(butylene terephthalate) cross-links rather well upon irradiation, especially with the use of a prorad (TAIC); the cross-linked material has a sufficient short-term heat resistance (260°C for 60 s), which makes it suitable to be used in lead-free soldering.

Polycarbonates (unstabilized grades) tend to discolor but are relatively resistant to irradiation, showing retention of elongation at yield and tensile modulus after irradiation up to 1000 kGy. The radiation resistance is the result of the presence of aromatic rings being in the molecule [51].

5.1.6.2 High-Performance Engineering Plastics

High-performance engineering plastics are suitable for service temperatures typically higher than 150°C (302°F), possessing exceptional mechanical properties. This group of polymers include polyimides (PI), poly(etherether ketone) (PEEK), poly(arylether sulfone) (PES), and poly(phenylene sulfide). When irradiated in a vacuum, many of these polymers are very stable and can show no change in physical and mechanical properties even after high beam doses. For example, aromatic PI have been shown to have resistance to both γ- and EB radiations up to doses of 100 MGy [52]. In the presence of oxygen, the physical and mechanical properties of these aromatic polymers can be dramatically changed. For example, an aromatic polysulfone showed no change in the flexural strength after irradiation with γ-rays to 6 MGy *in vacuo*. On the other hand, when the irradiation is carried out in the presence of air, the flexural strength drops to about half of its initial value at relatively low doses between 0.2 and 4 MGy [53].

The radiation for the series of PI, PEEK, and PES is shown to be excellent when compared to the related aliphatic polymers. The *G*-values for the evolution of gases were lower by between 0.01 and 0.0001 of the *G*-values for the corresponding aliphatic polymers. As for the resistance to EB irradiation, the order of these polymers is [54]:

$$\text{PI} > \text{PEEK} > \text{PES}$$

An increase of glass transition temperature occurs when either the crystalline or amorphous form of PEEK is irradiated. This is indicative that a cross-linking process is occurring [54, p. 2911].

The irradiation of amorphous and semicrystalline PPS with EB in the presence of nitrogen shows no noticeable change in mechanical or thermal properties at least to 10^4 kGy [55]. On the other hand, irradiation in air instead of nitrogen showed a change in both mechanical and thermal properties. At very high doses, i.e., 4×10^4 kGy, the amorphous PPS loses about 62% of its original tensile strength, while the semicrystalline PPS loses about 57%. The crystalline melting temperature (T_m) also changes, decreasing by about 10°C to 271°C.

5.2 EB Processing of Elastomers

Elastomers are polymeric substances with rubber-like behavior at ambient temperatures. This means they are more or less elastic, extensible, and flexible. They can be extended by relatively small force and return to the original length (or near it) after the force is removed. Rubber-like behavior can be observed in plastics, but under different conditions, such as at elevated temperatures or in swollen state. These materials are not true elastomers, however.

The macromolecules of elastomers are very long and flexible, randomly agglomerated, and entangled; the entangled macromolecular chains produce mechanical knots. Secondary forces acting between molecules are about 100 times weaker than the primary bonds between the atoms connected within the molecule. The secondary forces are of physical nature and their strength decreases rapidly with increasing distances between the molecules and with increasing temperature. The arrangement of the elastomeric molecules (flexibility and coiling of the chains, relatively weak intermolecular forces, chain entangling, and mechanical knots) gives elastomers the ability to undergo reversible deformations, but only under certain conditions: the deformation should not be too large and should last only a short time and within a narrow temperature interval. At higher deformations, particularly at elevated temperatures, the chains start to slide, the entanglements are reduced, and a permanent deformation takes place. The time dependence of these changes indicates a viscoelastic behavior of such polymeric material. The mechanical knots formed by the chain entanglements are not permanent and the weak intermolecular forces cannot guarantee a sufficient shape stability of the material, because they are affected by the changes of physical conditions (temperature, swelling) and they come into play only at low temperatures. Such material has only limited technical applications, because it has poor mechanical properties and sensitivity to temperature changes. It is predominantly plastic and dissolves in some liquids to give colloidal solutions.

Only when chemical bonds between neighboring molecules are introduced, is a raw elastomer converted into a rubber *vulcanizate*, which is essentially a three-dimensional network structure (Figure 5.1). The process is referred to as vulcanization or curing or more accurately as cross-linking. Cross-linked elastomer, or rubber vulcanizate, is capable of large reversible deformations within a broad temperature range and does not dissolve, but only swells in solvents and other liquids.

There are a large number of elastomeric materials used in commercial applications. They differ in the chemical nature of the monomer units, in their mutual arrangement and bonding, and in molecular weight, molecular weight distribution, branching, gel amount, etc. All these factors determine their chemical and physical properties, processing behavior, and solubility. Practically all commercial elastomers have the glass transition temperature below 0°C (32°F) and their molecular weights can range from 10,000 to 1,000,000 [56].

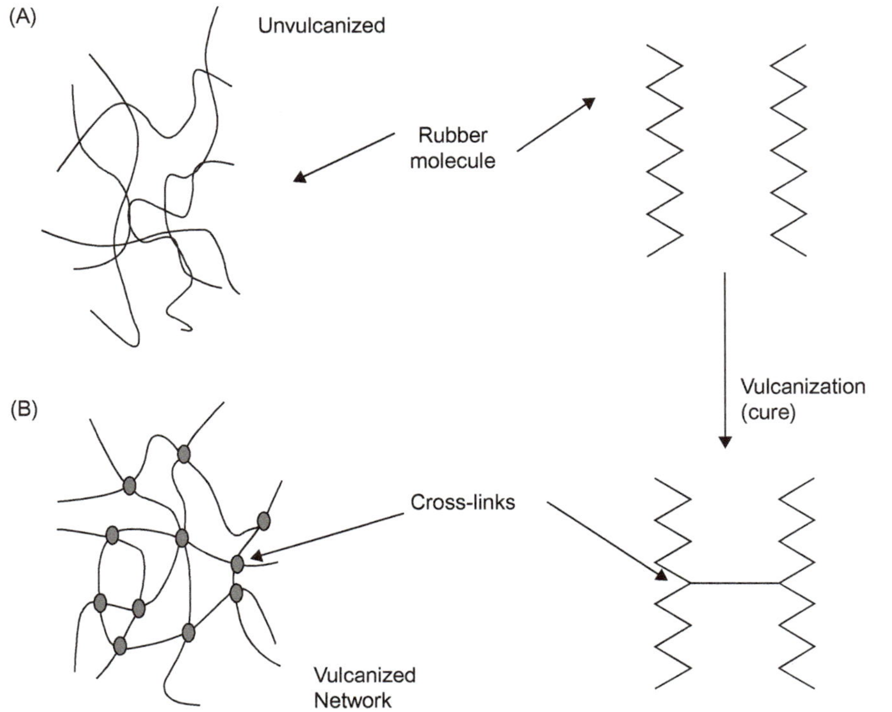

Figure 5.1 Elastomeric Material (A) Uncross-linked and (B) Cross-linked.

Vulcanization or cross-linking of elastomers is technically the most important process for conventional elastomers. During this process, strong chemical bonds are formed between molecules, restraining their mobility. As pointed out earlier, a three-dimensional network is formed. The cross-linking of elastomeric molecules is a random process; typically one cross-link is formed per 100–200 monomeric units.

The following changes occur as a result of cross-linking:

- The material changes from plastic to elastic state.
- The product is much less sensitive to temperature changes.
- The material becomes stronger and often harder.
- The shape fixed by vulcanization (e.g., in a compression mold) cannot be changed unless it is subjected to mechanical work.

There are several methods of vulcanizing standard commercial elastomers. The classic method involves sulfur or sulfur-bearing compounds and peroxides, and these are typically used for hydrocarbon elastomers, such as natural rubber (NR), styrene-butadiene rubber (SBR), polybutadiene rubber (BR),

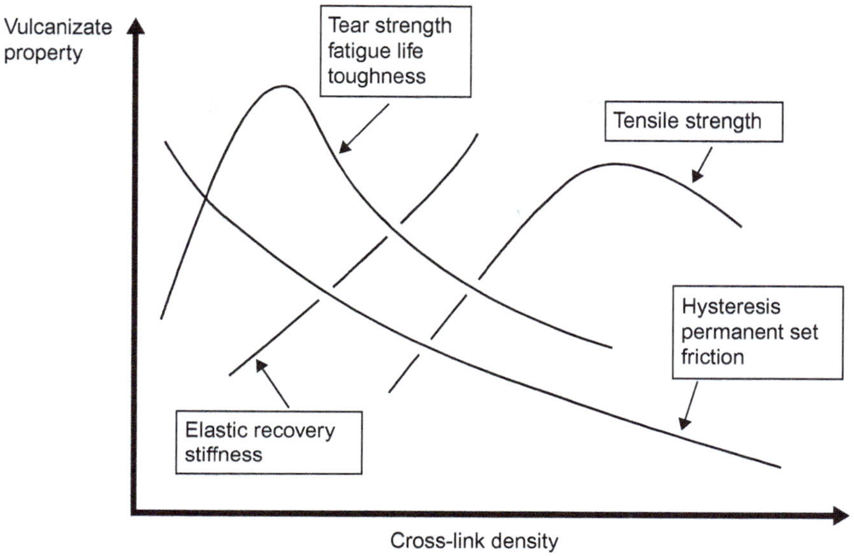

Figure 5.2 Effect of Cross-link Density on Selected Properties of an Elastic Material.

ethylene propylene rubber (EPM), and ethylene propylene diene rubber (EPDM). Other elastomers, such as chlorobutadiene rubber (CR), other chlorine-bearing elastomers, polyurethanes, and fluorinated rubber, use specific curing systems. Details about curing methods, compounding, and processing are covered in detail in several monographs [57–60].

Regardless of the method of cross-linking, mechanical properties of a cross-linked elastomer depend on cross-link density. Modulus and hardness increase monotonically with cross-link density and at the same time the network becomes more elastic. Fracture properties, i.e., tensile and tear strength, pass through a maximum as the cross-link density increases (Figure 5.2).

The cross-link density can be determined by equilibrium swelling or from equilibrium stress–strain measurements.

As pointed out earlier, an elastomer cross-linked above its gel point will not dissolve in a solvent but will absorb it and swell. The swelling will continue until the forces of swelling balance the retractive forces of the extended chains of the network. The cross-link density can then be calculated from the degree of swelling using the *Flory–Rehner equation*:

$$N = \frac{1}{2V_s} \frac{\ln(1-\phi) + \phi + \chi\phi^2}{\phi^{1/3} - \phi/2} \tag{5.1}$$

where N represents the moles of cross-links per unit volume (cross-link density), V_s is the molar volume of the solvent, ϕ is the volume fraction of polymer in the swollen gel, and χ is the polymer–solvent interaction parameter.

When using equilibrium stress–strain measurements, the cross-link density is determined from the *Mooney–Rivlin equation*:

$$\sigma/2(\lambda - \lambda^{-2}) = C_1 + C_2/\lambda \tag{5.2}$$

where σ is the engineering stress (force per unit of original cross-sectional area), λ is the extension ratio (measured length to original length), and C_1, C_2 are constants.

The value of C_1 is obtained from the plot of $\sigma/2(\lambda - \lambda^{-2})$ versus $1/\lambda$ and extrapolating to $1/\lambda = 0$. By comparison with the theory of elasticity, it has been proposed that $C_1 = 1/2NRT$, where N is the cross-link density, R is the gas constant, and T is the absolute temperature (of the measurement). To assure near-equilibrium response, stress–strain measurements are carried out at low strain rate and elevated temperature, and sometimes in the swollen state [61].

5.2.1 Physical Properties of Radiation Cross-linked Elastomers

Identically to chemically cross-linked (vulcanized) elastomers, the modulus of radiation cured gum elastomers depends on the concentration of elastically effective network strands and temperature [62].

Generally, values of tensile strength of radiation cured elastomers are reported to be lower than those of sulfur cured rubber [62, p. 610]. It can be shown that main chain scissions, prevalent particularly under certain irradiation conditions, have a distinct effect on the strength of cured elastomers. The effect can be understood in terms of reduction in molecular weight and an associated increase in the chain-end contribution [63]. If a certain modulus is to be achieved in a vulcanizate, regardless of whether or not scission is occurring, the increase in chain ends (chains which have no load-bearing capacity) has to be compensated for by an increase in the number of chemical cross-links. The associated decrease in the molecular weight of an elastically active strand (M_c) and its effect on failure properties such as energy to break is discussed in Ref. [63]. Since the maximum observable extension ratio $[(l/l_o)_{max}]$ is proportional to M_c, the increase in the ratio $G(S)/G(X)$ encountered during radiation cure should decrease the $[(l/l_o)_{max}]$ and consequently cause lower failure properties, such as tensile strength and energy to break [64,65].

There are certain polymers, such as polyisobutylene, which by nature of their microstructure have a high yield of radiation-induced scissions, regardless of how they were irradiated. However, even polymers, for which *G*(*S*) is negligible, when irradiated *in vacuo*, will undergo main chain scissions if the exposure is carried out at high temperature, in the presence of oxygen and/or certain additives. Degradation can also become significant with polymers having a low yield of cross-links *G*(*X*), since in the presence of a modest scission rate, the ratio *G*(*S*)/*G*(*X*) can be of significant magnitude [62, p. 611]. The compound ingredients such as antioxidants and oils may retard the rate of cross-linking of an elastomeric compound. Thus, the presence of cure retarding impurities and the exposure in an atmosphere containing oxygen may significantly contribute to the lower strength of radiation-cured elastomers. Another factor may be degradation of the polymer by ozone generated in the equipment during irradiation.

Obviously, there are many subtle differences in the structure, morphology, or network topology between radiation-cured and sulfur-cured elastomers [62, p. 612], but their physical properties may be nearly equal, provided that precautions are taken to avoid the occurrence of chain scissions. A comparison of radiation cross-linked and sulfur-cured NR (gum and carbon-black-reinforced compounds) is given in Table 5.9 [66,67].

Table 5.9 Comparison of Stress–Strain Properties of NR Cross-linked by Sulfur and Radiation

Gum (Compound Without Fillers)	**Sulfur Cured**	**Radiation Cured**
Tensile strength, MPa (psi)	27.8 (4026)	18.6 (2700)
	700	760
Elongation at break, % 300% modulus, MPa (psi)	2.3 (332)	1.7 (250)
Reinforced by 50 phr HAF Carbon Black	**Sulfur Cured**	**Radiation Cured**
Tensile strength, MPa (psi) Elongation at break, %	27.4 (3975)	22.8 (3300)
	470	350
300% modulus, MPa (psi)	14.8 (21.50)	17.9 (2600)

5.2.2 Effects of Radiation on Individual Elastomers

5.2.2.1 NR and Synthetic Polyisoprene

NR and guttapercha consist essentially of polyisoprene of *cis*-1,4 and *trans*-1,4 isomers, respectively. Commercially produced synthetic polyisoprenes have more or less identical structure but reduced chain regularity, although some may contain certain proportions of 1,2- and 3,4-isomers. Microstructure differences not only cause the polymers to have different physical properties but also affect their response to radiation. The most apparent change in microstructure on irradiation is the decrease of unsaturation. It is further promoted by the addition of thiols and other compounds [68]. On the other hand, antioxidants and sulfur were found to reduce the rate of decay of unsaturation [69]. A significant loss in unsaturation was found particularly in polyisoprenes composed primarily of 1,2 and 3,4 isomers [70,71].

A very important process occurring during irradiation is the formation of free radicals that leads to other changes, such as cross-linking, polymerization reactions, chain scission, and grafting.

About 98% of the gas formed during radiolysis of NR and polyisoprene is hydrogen, the rest consisting of methane and higher molecular weight hydrocarbons. The yield of hydrogen is directly proportional to dose, up to 2000 kGy (200 Mrad). Moreover, it is independent of dose rate and the type of radiation (γ, EB) [72].

Cross-linking of NR and synthetic polyisoprene has been studied by several investigators using EB and γ-radiation [73–75]. The general conclusion is that the yield of chemical cross-links $G(X)$ is constant with dose and independent of dose rate and the type of radiation used [75].

Effects of temperature are reported in Ref. [62, p. 617]. The value of the G-factor for cross-link yield, $G(X)$, measured at room temperature exposure, was 0.9. The ratio of chain scission to cross-links was found to be 0.05. Lowering the irradiation temperature to 77 K (−196.15°C or −321°F) had only a minor effect on $G(X)$ but markedly increased the yield of scissions $[G(S)/G(X)]$ to about 0.16. A more detailed analysis of the effect of temperature on radiation cross-linking of NR and a number of synthetic elastomers was published by Kozlov et al. [76].

The yield of cross-links in NR and in high *cis*-polyisoprenes was found to be about equal; however, polyisoprenes with high 1,2 and 3,4 configurations exhibited exceptionally high $G(X)$. The presence of 1,4 units contributed to the occurrence of chain scissions [76,77]. Oxygen has been found to increase the rate of scission and to decrease the degree of cross-linking [72,78]. Another interesting finding was that the rate of cross-linking is enhanced in

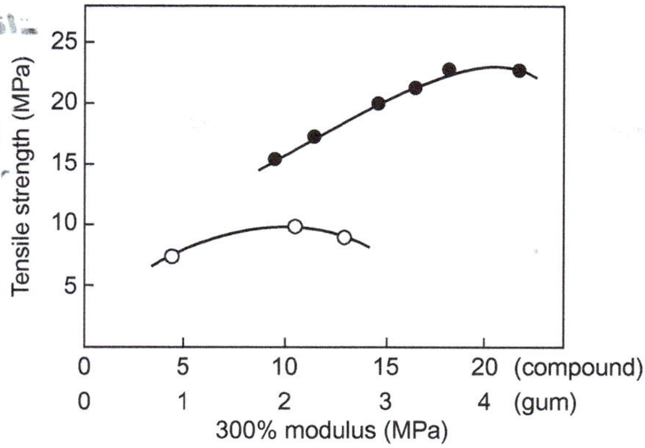

Figure 5.3 Tensile Strength of Irradiation Cured Purified NR. ○ gum, ● compound (50 phr N330 carbon black) [62]. *(Reprinted with permission from Rubber Division, ACS.)*

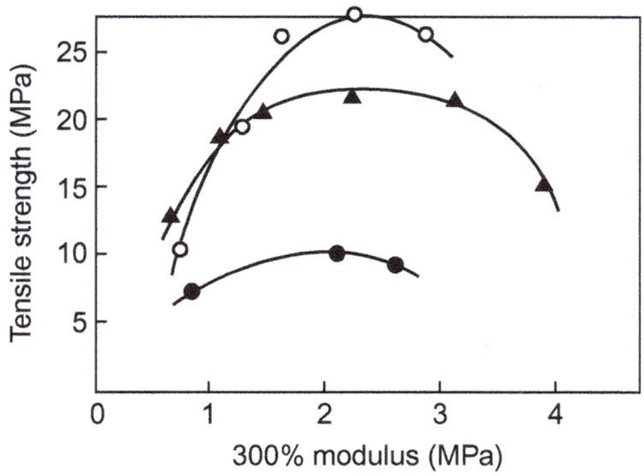

Figure 5.4 Tensile Strength of Radiation Cured Purified NR. ○ sulfur; ▲ peroxide; ● EB irradiation in nitrogen at 2.5 kGy/s [62]. *(Reprinted with permission from Rubber Division, ACS.)*

crystalline domains produced by strain orientation [78]. Maleimides and some halogenated compounds enhance the cross-linking of NR.

The most commonly reported physical properties of radiation cross-linked NR and compounds made from it are modulus and tensile strength, obtained from stress–strain measurements. Figure 5.3 illustrates some of the results obtained from gum rubber and from a NR compound reinforced by HAF carbon black [62 p. 619]. In Figure 5.4, the tensile strength of radiation cured gum is compared to that of vulcanizates cured by sulfur and peroxide.

Clearly, large doses are required for a full cure. The dose levels for attaining a maximum tensile strength are within the 200–500 kGy (20–50 Mrad) range. After exposure to such high levels of radiation, considerable modifications of NR by isomerization and chain scission processes occur [79]. Because of that, the maximum strength values of gum rubber achieved by ionizing radiation are lower than those achieved by sulfur- and peroxide-curing systems. When prorads (cross-link promoters), such as dichlorobenzene, maleimides, or acrylates are used, the curing dose is reduced and higher tensile strength values are found. Tensile strength data from radiation-cured NR are given in Table 5.10.

Somewhat greater tensile strength and elongation at break are obtained with high-energy electrons as compared with γ-radiation, when exposure is carried out in air. However, when antioxidants are added to the compound, the differences diminish [79,80].

The retention of the maximum tensile strength at elevated temperatures is greater for radiation cured than for chemically cured NR [21,80,81]. The physical properties after high-temperature aging are not improved, however. Lower flex life and higher abrasion resistance of radiation cross-linked NR have been reported [82]. Other properties such as permanent set, hardness, and resilience were found to be nearly equal.

The prevulcanization of NR in *latex form* has also been a subject of much investigation [83]. The cross-linking mechanism is not yet fully understood, but water apparently plays a major role in it. Irradiation results in the cross-linking of the rubber molecules and in coarsening of the latex particles. A process of cross-linking of NR latex has been developed to the point that it can be used for an industrial scale application [83]. The irradiation is performed in aqueous media by EB without a prorad ("sensitizer") at a dose of 200 kGy (20 Mrad) or in the presence of *n*-butyl acrylate at considerably lower doses, typically 15 kGy. The cross-linked film exhibits physical properties comparable to those obtained from sulfur-cured (vulcanized) film. As an alternative, the addition of a variety of chloroalkanes makes it possible to achieve a maximum tensile strength with radiation doses of less than 50 kGy (5 Mrad) [83].

5.2.2.2 Polybutadiene and its Copolymers

Homopolymers of polybutadiene can consist of three basic isomeric forms (*cis*-1,4, *trans*-1,4, and 1,2 vinyl), and these can be present in different sequential order. Copolymers may obtain a variety of comonomers, such as styrene and acrylonitrile. Depending on their distribution in the chain, random copolymers or block copolymers of different types and perfection can be produced. There are many synthetic elastomers based on butadiene available commercially.

Table 5.10 Tensile Strength Data from EB Cured NR

Compound	Amount of Additive (phr)	Dose (kGy)	Tensile Strength, MPa (psi) Reported	Note
Gum rubber (SMR-5L) plus *o*-dichlorobenzene	3	140	9.0 (1305)	15 kGy/s
Gum rubber (smoked sheets)	–	600	4.8 (696)	Irradiated in air
Smoked sheets and N330 carbon black	50	500	19.6 (2842)	Irradiated in air
		140	13.3 (1928)	
Smoked sheets and N330 carbon black plus octyl acrylate	60 20	140	18.3 (2654)	Irradiated in air
Smoked sheets and N330 carbon black plus tetramethylene diacrylate	55 10	140	18 (2291)	Irradiated in air
Smoked sheets and N330 carbon black plus glyceryl triacrylate	50 5	130	16.6 (2407)	Irradiated in air
Pale Crepe and N330 carbon black	50	160	15.8 (2291)	Irradiated in air
Pale Crepe and N330 carbon black plus octyl acrylate	50 20	160	16.6 (2407)	Irradiated in air

Upon irradiation, 1,4 polybutadienes and poly (butadiene–styrene) form free radicals relatively readily and their concentration has been found to increase linearly proportional to the dose, up to approximately 1000 kGy (100 Mrad) [62, p. 620].

During radiolysis of polybutadiene and butadiene–styrene copolymers, hydrogen and methane evolve [62,68,84–89]. The incorporation of styrene as comonomer strongly reduces the total gas yield. Small amounts (typically 2 phr) of *N*-phenyl-β-naphthylamine greatly reduce the gas yield [84] and at the same time reduce *G*(*X*) considerably.

Similarly to NR, polybutadiene and its copolymers exhibit decay of unsaturation during radiolysis. Its extent depends greatly on microstructure [62, p. 622] [89]. Another effect observed is *cis–trans* isomerization [89,90].

The yield of cross-links depends on the microstructure and purity of the polymer as well as whether it was irradiated in air or *in vacuo* [81]. The rate of degradation was found to be essentially zero when polybutadiene or poly (butadiene–styrene) were irradiated *in vacuo* but increased somewhat when irradiated in air.

The cross-linking of polybutadiene and its copolymers can be greatly enhanced by the addition of direct or indirect cross-link promoters. Largest *G*(*X*) values are obtained when using polymercaptans in combination with halogenated aromatics [87,91,92]. Carbon blacks and silica fillers were found to enhance cross-linking and, to some extent, become chemically linked to the polymer chains [64,93]. Some investigators suggest that EB irradiation of carbon-black-reinforced rubber compounds results in an increased cross-link density near the surface of the carbon black particle [94]. Aromatic oils enhance the occurrence of chain scissions and decrease the rate of cross-linking [95]. Common accelerators for rubber, such as tetramethylthiuram disulfide, sulfur, diphenylguanidine, and mercaptobenzothiazole, inhibit radiation cross-linking, the degree of retardation being in the order of their listing [95]. As pointed out earlier, several chemicals were reported to protect polybutadiene and its copolymers against radiation damage.

The reactions leading to the formation of cross-links between polymeric chains also can be employed to attach various compounds to the polymeric backbone. Using this technique, grafting of styrene [62, p.629] [96], mercaptans [62, p. 629], carbon tetrachloride [62, p. 629], [97] and other compounds can be carried out.

The information on physical properties of radiation cross-linking of polybutadiene rubber and of butadiene copolymers was obtained in a similar fashion as for NR, namely by stress–strain measurements. From Table 5.11, it is evident that the dose required for a full cure of these elastomers is lower than for NR. The addition of prorads allows further reduction of the cure

Table 5.11 Tensile Strength Data from EB Cross-linked Polybutadiene and its Copolymers

Polymer	Comonomer (mol%)	Additive	Amount of Additive(s) (phr)	Dose (kGy)	Tensile Strength, MPa (psi) Reported
Polybutadiene	0	N330 carbon black and *p*-dichlorobenzene	50 2	80	15.2 (2204)
Poly (butadiene–styrene)	22.5	N330 carbon black and *p*-dichlorobenzene	50 2	100	23.45 (3400)
	23.5	N330 carbon black	50	280	20.4 (2958)
	25	N330 carbon black	50	400	19.2 (2784)
		N330 carbon black plus hexachloroethane	50 3	150	18.1 (2624)
Poly (butadiene–acrylonitrile)	32	Semi-reinforcing carbon black	50	100	22 (3190)

dose with the actual value depending on the microstructure and macrostructure of the polymer and also on the type and concentration of the compounding ingredients, such as oils, processing aids, and antioxidants in the compound. For example, solution-polymerized polybutadiene rubber usually requires lower doses than emulsion-polymerized rubber because it contains smaller amount of impurities than the latter. Since the yield of scission $G(S)$ is relatively small, particularly when oxygen is excluded, tensile strengths comparable to those attainable with standard sulfur-cured systems are obtained [98]. Radiation-cured polybutadiene and its copolymers were reported to have superior tensile strength and ultimate elongation at high temperatures [99]. The abrasion resistance was found to be higher; most of the higher properties, such as hardness, resilience, and permanent set, were equal to those of chemically cured compounds of this category. Tensile strength values near equal to those obtained by chemically cured NBR were reported [64,82]. Relatively low-radiation doses were required for a complete cure of these materials. Radiation-cured carbon–black-reinforced SBR was found to have somewhat lower tensile strength and a lower maximum extensibility than sulfur-cured compounds, considerably longer fatigue flex life, and slower crack growth [81]. Recent reports indicate a considerable improvement in methods for cross-linking of SBR by using 1–3 phr of di- or triacetylate prorad in a compound with 45 phr reinforcing carbon black (e.g., HAF). The appropriate dose for this material is 200 kGy. The product has better mechanical properties and thermal stability than a similar chemically cross-linked SBR compound [18, p. 107].

5.2.2.3 Polyisobutylene and its Copolymers

To this category belong homopolymers of polyisobutylene, copolymers of isobutylene and isoprene (butyl rubber), chlorobutyl, and bromobutyl. All these have been produced commercially for decades.

Upon irradiation, polyisobutylene and its copolymers tend to degrade. There are a large number of studies determining the nature of the process and mechanisms and it was established that it involves formation and reaction of free radicals [100]. The radical concentration increases linearly with dose up to 1000 kGy (100 Mrad) [101].

Hydrogen and methane contribute to about 95% of the gas yield during radiolysis of polyisobutylene. The rest is made up of isobutylene and other mostly unidentified fragments [62, p. 635] [102–105]. It was found that the ratio of methane to hydrogen remains constant over a wide range of dose, whereas the ratio of isobutylene to hydrogen or methane rises rapidly with dose [102,103]. Moreover, it has been reported that the gas yields are largely independent of irradiation temperature [106].

The main processes observed and reported are formation of vinylidene double bonds [62, p. 637] [102,103] and abstraction of methyl groups. The concentration of unsaturations located mainly at the end of a polymer chain produced during radiolysis was found to increase with dose [106]. Moreover, a linear relationship between unsaturations and scissions formed was noted over the temperature range of 83–363 K (−310 to 194°F). The ratio of double bonds to scissions was found to be independent of temperature and dose [102,103,105].

Degradation is the predominant process in irradiation of polyisobutylene. Generally, the scission yield increases with temperature in accord with the temperature dependence of the formation of unsaturation [107]. For poly (isobutylene–isoprene), it was found that the rate of degradation decreased with the increased content of isoprene in the copolymer, in accord with peroxide studies [108]. An extrapolation of this function to higher concentrations of unsaturation indicated that net cross-linking is expected to occur above approximately 5 mol% isoprene. A drastically different response to radiation was found for chlorinated poly(isobutylene–isoprene), poly(isobutylene–isoprene–divinylbenzene) terpolymers [87] and dehydrohalogenated chlorobutyl rubber [62, p. 638]. Rapid gelation occurred in those polymers already at low doses. The gradual depletion of these reactive sites by cross-linking then leads to prevailing degradation [109]. Certain additives can affect the response of these elastomers to irradiation. Polymercaptan compounds retard the net degradation of polyisobutylene and cause rapid gelation of poly(isobutylene–isoprene) and chlorobutyl rubber [87,91,92]. Allyl acrylate was found to cause gelation of polyisobutylene [110].

It is possible to cross-link polyisobutylene and poly (isobutylene–isoprene) and chlorobutyl rubber by means of certain additives; however, the physical properties of the vulcanizates prepared by conventional curing methods cannot be attained. The most promising results were obtained with chlorobutyl compounds containing thioether polythiols as prorads [92].

5.2.2.4 Ethylene–Propylene Copolymers and Terpolymers

Commercial grades of ethylene–propylene copolymers (EPR) contain 60–75 mol% of ethylene to minimize crystallization. The addition of a third monomer, such as 1,4 hexadiene, dicyclopentadiene, or 5-ethylidene-2-norbornene, produces generally amorphous faster curing elastomers. A large number of such terpolymers, referred to as EPDM, are available commercially. Their properties, performance, and also their response to radiation vary considerably depending on macrostructure and ethylene/propylene ratio, as well as on the type, amount, and distribution of the third monomer. When EPDM is irradiated, the nature of the radical formed depends on the third monomer used [62, p. 640] [111] and as in most cases, the radical concentration increases linearly with dose [112].

Cross-linking rate of EPR by radiation comes close to that of PP. EPDM terpolymers exhibit enhanced cross-linking rate and it increases with the diene content. However, not only cross-linking rate but also a greater yield of scissions results from the addition of the third monomer [113].

Cross-linking of EPR can be promoted by the addition of a variety of additives, particularly by those which were found effective in PP. Tetravinyl silane, chlorobenzene, nitrous oxide, allyl acrylate, neopentyl chloride [114,115], and *N*-phenyl maleimide [116] were reported to promote the process.

Radiation-cured carbon-black-filled EPDM compounds exhibit tensile strength comparable to that of their analogues vulcanized by sulfur-accelerator systems [62, p. 642]. Similar results were obtained from compounds containing up to 20 phr of acrylic additives, such as trimethylol propane trimethacrylate, ethylene glycol dimethacrylate, triallyl phosphite, and TAC [62, p. 642]. It should be noted, however, that some of the resulting compounds may be composed of two more or less independent, at least partially, separated networks. The noted lower compression set and reduced swelling in oils of the radiation-cured samples most likely result from this different morphology [117]. Among the variety of EPDM grades, those containing ethylidene norbornene exhibit the fastest rate of radiation cross-linking [116–118].

Extender oils were found to cause a considerable increase in the dose required to attain the optimum cure. This can be explained by reaction of transient intermediates formed on the irradiated polymer chain with the oil and with the energy transfer, which is particularly effective when the oil contains aromatic groups. Thus, the ranking of oils as to their cure inhibition is aromatic > naphthenic > aliphatic [117]. This aspect is very important because many carbon-black-reinforced EPDM compounds contain frequently 100 phr or more oil.

5.2.2.5 Polychloroprene

Polychloroprene or "Neoprene" (CR) is the polymer of 2-chloro-1,3 butadiene. Emulsion polymerization produces an almost entirely *trans*-1,4 polymer, which is highly crystalline. In the case of such crystallizable CR, the cross-link density has been found to be lower in the crystalline region as compared to the amorphous region due to the lower radical mobility in the former. Less crystalline CRs are produced by incorporating several weight % of 2,3-dichloro-1,3 butadiene into the polymer to break up crystalline sequences.

Irradiation of carbon-black-reinforced CR compounds produced maximum tensile strength of 20 MPa (2900 psi) at a dose of 200 kGy (20 Mrad), which is a value obtained typically from chemically cured compounds. An addition of 20 phr of *N*, *N′*-hexamethylene-bis methacrylamide as a prorad in the above compound produced a tensile strength of 18 MPa (2610 psi) at a dose of 70 kGy (7 Mrad). Further addition of 6 phr of hexachloroethane caused the deterioration of the tensile strength by 50% at the 70 kGy (7 Mrad) dose [119].

When irradiating a 1:1 blend of CR and poly (butadiene–acrylonitrile) (NBR) reinforced by 50 phr furnace black and containing 5–15 phr of tetra methacrylate of bisglycerol phthalate, it exhibited a tensile strength of 20 MPa (2900 psi) at a dose of 150 kGy (15 Mrad) with values of elongation at break in the range 420–480%. These values are equal to or better than those obtained from similar compounds cured chemically [120].

Irradiation of CR latexes of two different structures, one containing some sulfur and having a lower degree of branching and another a highly branched polymer, made by mercaptan modification, showed a more rapid cross-linking of the branched polymer. The presence of the latex dispersion medium further enhanced the cross-linking process. The concentration of free radicals in the irradiated latexes was about 50% higher than that in the coagulated and subsequently dried rubber films [121].

Polyorthoaminophenol (2 wt%) and polyorthoaminophenol + phenyl-β-naphthylamine (0.7 wt% of each) acted as antirads, preventing any significant cross-linking on exposure to 240 Gy (24 krad) [122,123]. The addition of a diphenylamine derivative gave a reasonable protection to radiation up to 220 kGy (22 Mrad) [123].

The addition of polyfunctional monomers, such as PE glycol dimethacrylate, enhances the degree of cross-linking of the material. The degree of cross-linking has been found to be proportional to the absorbed dose and the initial concentration of the polyfunctional monomer [124].

5.2.2.6 Nitrile Rubber

Nitrile rubber (NBR) is available in three grades with three levels of acrylonitrile, namely 18%, 33%, and 50%. Properly compounded nitrile rubber can be cross-linked efficiently using diethyleneglycol dimethacrylate at 20 kGy with results similar to those obtained from conventionally cured NBR [125]. Maximum strength of materials based on all three grades is obtained at a dose of 50 kGy. There is a drop in tensile strength at higher absorbed doses [126].

5.2.2.7 Silicone Elastomers

There is a large variety of elastomers based on the –Si–O–Si– chain, with different groups attached to the Si atom, which affects the properties of the polymer. Technologically, the most widely used silicone elastomers are those with all methyls on the silicone atoms, i.e., polydimethylsiloxanes (PDMS) or ones with less than 0.5 mol% of vinyl substitution for the methyls [126].

Irradiation of PDMS produces hydrogen, methane, and ethane. The gas yield at room temperature correlates with the concentration of cross-links formed [127]. This can be expected since double bonds cannot be formed.

Cross-linking is the predominant process during irradiation of siloxane polymers. Chain scissions are negligible [128–130]. The cross-link density increases linearly with dose up to 1600 kGy (160 Mrad) [130]. At 5.0 MGy (500 Mrad), the $G(X)$ value is 0.5 [131]. Free-radical scavengers, such as *n*-butyl and *tert*-dodecyl mercaptan, and diethyl disulfides, are the most effective antirads [132,133]. At concentration of 10%, two-thirds of the cross-links were prevented from forming; however, the scission yield was also increased.

The tensile strength of PDMS irradiated by a dose of 60 kGy (6 Mrad) is 15% lower than that of PDMS cross-linked by peroxide, but when 0.14 mol% of vinyl unsaturation is substituted for the methyl groups, the tensile strength is 30% higher [132]. Irradiation of PDMS containing 55 phr silica filler using a dose of 40 kGy exhibited a superior abrasion resistance when compared to peroxide vulcanizates even after prolonged thermo-oxidative aging at 523 K (482°F). The tensile properties of the materials prepared by both these cross-linking methods were similar. An optimum tensile strength of 8 MPa (1160 psi) was obtained on exposure to 40 kGy (4 Mrad) [131].

5.2.2.8 Fluorocarbon Elastomers

FKMs represent the largest group of fluoroelastomers. They have carbon-to-carbon linkages in the polymer backbone and a varied amount of fluorine in the molecule. In general, they may consist of several types of monomers: PVDF, HFP, trifluorochloroethylene (CTFE), PTFE, perfluoromethylvinyl ether (PMVE), ethylene or propylene [134]. Other types may contain other comonomers, e.g., 1,2,3,3,3-pentafluropropylene instead of HFP [135]. FKM exhibit good chemical and thermal stability and good resistance to oxidation.

Since fluorocarbon elastomers being discussed here contain hydrogen in their molecules, they have the tendency to cross-link in addition to undergoing scission, common in fluoropolymers when exposed to radiation. The cross-linking predominates, but there is still a significant degree of chain scission [36, p. 339] [62, p. 649].

The use of cross-linking promoters (prorads), such as diallyl maleate, TAC, TAIC, trimethylolpropane methacrylate (TMPTM), and *N,N'*-(*m*-phenylene)-bismaleimide (MPBM) at amounts up to 10 wt%, reduces the dose needed and the damage to the elastomeric chain by the radiation. It appears that individual FKM have the best cross-link yield with a specific prorad [36, p. 339] [136,137,138]. For example, copolymers of vinylidene fluoride (VDF) and HFP, e.g., Viton® A and copolymers of VDF and chlorotrifluoroethylene (Kel-F®), attain a high gel content and thermal stability up to 473 K (392°F) when cross-linked with a dose as low as 10–15 kGy without prorads and only 2 kGy with 10 phr of multifunctional acrylates. Such thermal stability is unattainable with conventional chemical curing. Mechanical properties (tear resistance, tensile strength) are better than those from vulcanizates prepared by peroxide cross-linking [136]. The VDF/HFP copolymer is cross-linked by EB more efficiently than the copolymer of VDF and CTFE [22, p. 439].

Tetrafluoroethylene-propylene (TFE-P) copolymers can be cross-linked effectively by the addition of allyl compound, such as diallyl maleate [138, 139]. TFE-P compounds containing 10–60 phr (parts per hundred parts of rubber) were cross-linked by irradiation by EB up to dose of 1000 kGy [139]. The $G(X)$ and $G(S)$ of this polymer are estimated to be 0.86 and 0.13, respectively [140]. The advantage of radiation cross-linking of this elastomer over a chemical cure is the absence of contaminants, which is of extreme importance in the semiconductor industry.

Perfluoroelastomers (ASTM designation FFKM) are copolymers of two perfluorinated monomers, such as TFE and PMVE with a cure site monomer (CSM), which is essential for cross-linking. Certain FFKMs can be cross-linked by ionizing radiation [141,142]. The advantage of radiation-cured FFKM is the absence of any additives, so that the product is very pure. The disadvantage is the relatively low upper-use temperature of the cured material, typically 150°C, which limits the material to special sealing applications only [141].

5.2.2.9 Fluorosilicone Elastomers

Fluorosilicone elastomers generally respond to ionizing radiation in a similar fashion to silicone elastomers (PDMS) [143]. One interesting application is a process of preparing blends of fluoroplastics, such as PVDF with fluorosilicone elastomers, to obtain materials having a unique combination of flexibility at low temperatures and high mechanical strength [142].

Efficiencies of radiation-induced reactions of selected elastomers are given in Table 5.12. Tensile strength data from another group of elastomers are given in Table 5.13.

5.2.2.10 Thermoplastic Elastomers

Thermoplastic elastomers (TPEs) are either block copolymers (SBS, SEBS, SEPS, TPU, COPA, COPE) or blends, such as TPO (elastomer/hard

Table 5.12 Efficiencies of Radiation-Induced Elastomer Reactions

Polymer	*G*(*X*)	$G(H_2)$[a]	*G*(*S*)/*G* (*X*)[a]
Polyisoprene (NR)	0.9	0.43–0.67[b]	0.16
		0.25–0.3[c]	
Polybutadiene	3.6	0.23[d]	0.1–0.2
Styrene–butadiene copolymer with 60 mol% styrene	0.6	0.11[d]	–
Styrene–butadiene copolymer with 23.4 mol% styrene	1.8–3.8	0.45 (87% H_2)	0.2–0.5
Ethylene–propylene rubber	0.26–0.5	3.3	0.36–0.54
			G(*S*)= 0.3–0.46
EPDM (with dicyclopentadiene co-monomer)	0.91	–	*G*(*S*)= 0.29
Polyisobutylene	0.05	1.3–1.6	*G*(*S*)= 1.5–5
Copolymer of VDF and HFP	1.7	0.27 *G* (HF) = 1.2	*G*(*S*)= 1.36
Copolymer of VDF and chlorotrifluoroethylene	1.03	–	*G*(*S*)= 1.56

[a] *Unless otherwise stated.*
[b] *For doses up to 2 MGy (200 Mrad).*
[c] *Polyisoprenes with high 1,2 and 3,4 content.*
[d] *For a mixture of H_2 and CH_4.*

thermoplastic) and TPV (thermoplastic vulcanizate, blend of a vulcanized elastomer and a hard thermoplastic). These types represent the majority of the TPE; other types are either specialty or small volume materials.

Normally, TPEs are not cross-linked because their thermoplastic nature is the desired property in most cases. However, in some cases, cross-linking is used to improve mechanical properties, to influence flow [143] to reduce swelling in oils and solvents, to eliminate dissolution of the polymer in oils and solvents, and to increase heat resistance and influence other performance

Table 5.13 Tensile Strength Data from Selected EB-Irradiated Elastomers

Polymer	Additive	Amount of Additive (phr)	Dose (kGy)	Tensile Strength, MPa (psi)
Natural rubber	*o*-Dichlorobenzene	3	140	9 (1305)
	Mono- and multifunctional acrylates	5–20	140	15–18 (2175–2610)
	Carbon black	50–60	1000	1.6 (232)
	No additives	–	600	4.8 (696)
Polybutadiene	Carbon black and dichlorobenzene	50 2	8000	15.2 (2204)
Copolymer of butadiene and styrene (23.5 mol% styrene)	Carbon black	50	28,000	20.4 (2958)
NBR (copolymer of butadiene and acrylonitrile) (32 mol% acrylonitrile)	Carbon black	50	10,000	22 (3190)
Polychloroprene	Carbon black	NA	20,000	20 (2900)
	Carbon black and diacrylamide	20 NA	700	18 (2610)
Blend of CR and NBR	Carbon black and multifunctional acrylate	50 5–15	15,000	20 (2900)

Table 5.14 Effect of Radiation Dose on Performance Characteristics of a Copolyamide Formulation with 2% of Prorad

Property	Dose (kGy)				
	0	50	100	150	200
Tensile strength (MPa)	59.5	65.7	53.4	42.9	Broke
Elongation at break (%)	400	350	200	125	Broke
Hot creep elongation at 29 psi, 200°C (%)	Melted	55	61	58	63

Material copolyamide (COPA) Pebax®, hardness 72 Shore D.
Source: MD + DI (Medical Device and Diagnostic Industry on line, June 1, 1999, mddionline.com)

characteristics. Examples where cross-linking by ionizing radiation is necessary for the given processes are:

- Wire and cable insulation (to improve resistance to abrasion and cut-through)
- Thermoplastic foams where a partial cross-linking by EB increases melt strength
- Heat-shrinkable films, sheets, and tubing.

An example of cross-linking of COPA material by ionizing radiation to improve its heat resistance is given in Table 5.14.

TPEs based on polyolefins (TPO) are blends of PE or PP with EPDM elastomers wherein the elastomer is often cross-linked using thermochemical systems [144]. TPOs more suitable for medical products with no chemical residuals can be made using EB processing to cross-link the elastomer portion in such an elastomer-plastic blend. The thermoplastic governs the melt transition and thus the extrusion properties of TPOs. The radiation response of these materials is also governed by the choice of the thermoplastic. An example of an EB-cured blend of EPDM and PE is used for fluid transmission tubing and electrical insulation [145].

See the review article on EB processing of elastomers by Drobny [146].

5.3 EB Processing of Liquid Systems

EB processing of solvent-free liquid systems for coatings, inks, and paints involves essentially polymerization and cross-linking using electrons with

Table 5.15 Typical Dose Ranges for Cross-linking of Selected Elastomers

Polymer	Dose (kGy)
EPM	50–150
EPDM	100–150
CSM[a]	100–150
VMQ[b]	50–125

[a] *Chlorosulfonated PE "Hypalon."*
[b] *Vinyl–methyl silicone rubber.*

energies between 120 and 300 keV. Initiation by electrons leads primarily to free-radical reactions. Cationic polymerization is only found in rare cases [147–151].

Typical liquid systems require doses between 10 and 50 kGy. They consist of binders (prepolymers) with acrylic ($H_2C{=}CH{-}CO{-}O{-}$) double bonds in the main chain (polymaleates and polyfumarates) and of monomers, usually acrylates used as reactive thinners [152]. Other ingredients added to the formulation may be pigments, dyes, fillers, flatting agents, and additives to improve film and surface properties and attain the required performance criteria [153].

Reactive prepolymers used as binders are produced by acrylation of oligomers, such as epoxy resins, urethanes, polyesters, silicones, oligo-butadiene, melamine derivatives, cellulose, and starches [154–156]. Prepolymers are the principal ingredients of coating formulations and largely determine the basic properties of the coating. Examples of industrially important acrylated prepolymers are given in Table 5.15.

Monomers, also called reactive thinners, are used to reduce viscosity of the prepolymers but also have an effect on properties of the cured film. They form a high molecular weight network with the prepolymer after curing. In order to attain an adequate degree of cross-linking, principally bifunctional and polyfunctional acrylates are used. Monofunctional acrylates give a less reactive coating and are less desirable ingredients because of their volatility, odor, and skin-irritating effects. Currently, the following bifunctional and polyfunctional acrylates are used in industrial applications (abbreviations in parentheses):

Tripropylene glycol diacrylate	(TPGDA)
1,6-Hexanediol diacrylate	(HDDA)
Dipropylene glycol diacrylate	(DPGDA)

Trimethylolpropane triacrylate	(TMPTA)
Trimethylolpropane ethoxytriacrylate	(TMP(EO)TA)
Trimethylpropane propoxytriacrylate	(TMP(PO)TA)
Pentaterythritol triacrylate	(PETA)
Glyceryl propoxytriacrylate	(GPTA)

Ethoxylated and propoxylated acrylates are highly reactive, yet exhibit less skin irritation [151].

EB curing of coatings requires inert atmosphere to prevent oxygen inhibition. Inertization is done by nitrogen, obtained by vaporization of liquid nitrogen. Another advantage of using inert atmosphere is to prevent formation of ozone, which is formed in the presence of air. Often part of the gas is used for convective cooling of the accelerator beam window.

Many of the liquid systems curable by EB equipment are very similar to those cured by UV radiation. EB process has the advantage that it offers the possibility of curing much thicker coats and pigmented formulations at much higher speeds than UV radiation. There is no need to use photoinitiator systems, which are often expensive and sometimes discolor the finished film. EB-cured coatings have frequently better adhesion to substrate because of the penetration of the electrons. These advantages are offset by the much higher capital cost of the EB equipment [2]. Another issue is to have enough production to sufficiently utilize the highly productive EB equipment. In some cases, to combine UV and EB processes is of technological and/or economical advantage.

5.4 Grafting and Other Polymer Modifications

5.4.1 Grafting

The purpose of grafting is to add new properties or functionalities, such as changing wettability, adhesion, printability, metallization, anti-fog and anti-static properties, and biocompatibility to the polymer, in this context referred to as the *trunk polymer*. There are many methods to initiate the grafting polymerization, e.g., plasma treatment, ultraviolet light radiation, decomposition of initiators, oxidation polymers, enzylatic grafting, and high-energy irradiation (γ-rays, EB) irradiation. Among these methods, EB irradiation may be the convenient and most effective for industrial usage because of its high irradiation dose, ease of generating active sites in many kinds of polymers, effective penetration into the polymer, and moderate initiation reaction conditions (atmospheric pressure and room temperature). Since ionizing radiation readily generates free radicals, the majority of grafting reactions proceed by the free-radical mechanism.

There are essentially no restrictions on the kind or form of the trunk polymer for the radiation grafting. Radiation-induced grafting in its simplest form involves heterogeneous systems, with the substrate being film, fiber, or even powder with the monomer to be grafted onto the substrate being a neat liquid, vapor, or solution [18, p. 334] [157–160]. The resulting whole polymer is referred to as the *graft copolymer* and is written as P-g-M, where P is the trunk polymer and M is the monomer used for the grafting. The monomers used for grafting may be in the form of the gas, liquid, solution, or emulsion.

Currently, three main radiation grafting techniques are known [157,161,162]:

1. The preradiation method
2. The peroxidation method
3. Simultaneous (or direct) method.

In the two-step *pre-radiation procedure*, the substrate is first irradiated, usually *in vacuo* or inert gas, to produce relatively stable-free radicals, which are then reacted with a monomer, usually at elevated temperatures. This technique is suitable mainly for semicrystalline polymers that can trap radicals in the crystalline region [162]. A major advantage of the preradiation method is the minimization of homopolymer formation [162].

The *Peroxidation procedure*, which is the least often used of all the irradiation techniques, involves irradiation of the substrate in the presence of air or oxygen. This produces diperoxides and hydroperoxides on the surface of the substrate, which are stable and the substrate can be stored until combination with a monomer is possible. Monomer, with or without solvent, is then reacted with the activated peroxy trunk polymer in air or under vacuum at elevated temperatures to form the graft copolymer. The advantage of this method is the relatively long shelf life of the intermediate peroxy trunk polymers before the final grafting step [163].

In the *simultaneous method*, which is the one most commonly used, the substrate is irradiated while in direct contact with the monomer. The monomer can be present as a vapor, liquid, or solution. This grafting process can occur via a free-radical or ionic mechanism [162,164]. With the simultaneous method, the formation of homopolymer is unavoidable, but there are several systems to minimize this. The advantage of this method is that both monomer and substrate are exposed to the radiation source and both form reactive sites. The other two techniques rely upon rupture of bonds to form reactive sites and therefore require higher radiation doses. Thus, the simultaneous method is more suitable for substrates sensitive to radiation. The simultaneous method can utilize UV radiation besides the EB source. Logically, UV irradiation requires photoinitiator and/or sensitizer to achieve an acceptable level of grafting.

Radiation grafting can be performed with the monomers being neat or dissolved. In some cases, the use of solvents can produce graft copolymers with unique properties. Solvents, which wet and swell the backbone polymers, often assist grafting. Certain additives, including mineral acids and inorganic salts, such as lithium perchlorate, as well as monomers, such as DVB and TMPTA, improve grafting yields [162, p. 99].

The dose rate affects both the yield and chain length of the grafted material. Air has a detrimental effect on grafting since it inhibits the reaction, which is consistent with other radiation-induced free-radical reactions. Increasing the temperature of the grafting system increases the yield. This is very likely because raising the temperature increases the diffusion rate of the monomer into the substrate [163].

The efficiency of radiation energy in the direct method is higher than that in the postirradiation method. Doses of 10 kGy or less are enough for the direct method; however, the actual dose depends on the combination of the trunk polymer and the monomer used. For the two-step method, the dose may be as high as 100 kGy to achieve a sufficient degree of grafting. Postirradiation heating can improve the outcome of the grafting process [164].

Theoretically, radiation grafting is applicable to any organic backbone polymer, including PE, PP, fluoropolymers, PVC, cellulose, and wool. PVC grafting yields are among the highest under the radiation grafting conditions. Acid enhancement is observed for most backbone polymers. Solvent effects are also relevant to many backbone polymer systems (Table 5.16). At low styrene solution concentrations (e.g., 30%), a peak in grafting is observed. This, the *Trommsdorff peak* [165], is important since the chain length of graft is a maximum under these conditions and the graft yield reaches a maximum.

Diffusion of the monomer controls chain growth and chain termination in the internal structures of the polymer [166]. The solubility parameters δ of the solvent should be close to that of the polymeric backbone so that the necessary chemical energy to disrupt intermolecular cohesive forces between polymer chains and permit chain mobility, is created. An example is PET, an (AB) alternating copolymer where A is the semirigid aromatic segment with a δ-value of 9.8 and B is a flexible aliphatic ester with a value of 12.1. Solvents, such as dimethylsulfoxide (DMSO) (12.93), pyridine (10.61), and dichloroethene (DCE) (9.0) with solubility parameters close to PET, promote diffusion and incorporation of the monomer and subsequently grafting takes place [167]. Wetting of the polymer by the solvents is another important criterion; the surface tension data give useful information regarding its key role in grafting [167].

Table 5.16 Effect of Solvents on Radiation Grafting of Styrene to PP[a]

	Grafting Yield (%)							
	Styrene (wt% in Solution)							
Solvent	**20**	**30**	**40**	**50**	**60**	**70**	**80**	**100**
Methanol	29	94	50	37	36	35	29	22
Ethanol	44	89	65	47	36	32	30	22
n-Butanol	123	74	34	40	33	29	28	22
n-Octanol	49	107	68	42	32	29	26	22
Dimethylformamide	24	40	43	44	40	39	33	22
Dimethylsulfoxide	11	29	66	61	56	42	24	21
Acetone	13	20	24	25	22	24	25	22
1,4-Dioxane	6	12	15	17	19	21	23	22

[a] *Total dose 3 kGy, dose rate 0.4 kGy/h.*
Source: Ref. [162].

Table 5.17 Radiation Grafting of Miscellaneous Monomers[a] onto Cellulose

Monomer	Grafting yield (%)
Styrene[b]	40
o-Methylstyrene[b]	110
p-Methylstyrene[b]	6
o-Chlorostyrene[b]	74
2-Vinyl pyridine[b]	3
Methyl methacrylate[c]	18
Vinylacetate[d]	11

[a]40 vol.% in methanol.
[b]Total dose 5.4 kGy; dose rate 0.45 kGy/h.
[c]Total dose 10 kGy; dose rate 0.45 kGy/h.
[d]Total dose 10 kGy, dose rate 3.9 kGy/h.

Substituents have an effect as in conventional polymerization reactions. In radiation grafting, certain substituents activate monomers, while others deactivate them [168]. These effects can be seen in Table 5.17.

Multifunctional monomers, such as acrylates (e.g., TMPTA), were found to have dual function, namely to enhance the copolymerization and also to cross-link the grafted trunk polymer chains. The addition of an acid along with a polyfunctional monomer has synergistic effects on grafting [169].

Although many radiation-grafted materials have been discovered, only a limited number of them have been commercially utilized [170] [162, p. 112]. One of the first successful applications was the use of grafted films in battery separators [171]. Other applications are in ion exchange resins and membranes for separation processes [162, p. 114]. The textile industry represents opportunities in improving flame retardation, permanent press, dyeing, and antistatic properties [161,172]. Other processes based on radiation grafting are useful in medicine (diagnostic and therapeutic, biocompatible materials) and industry (fermentation, bioseparation), for producing catalyst supports [173] and in proton exchange membranes of fuel cells [174].

Radiation rapid curing (RRC) is another reported application of radiation grafting [162]. It is used for films containing oligomer/monomer mixtures, which cross-link within a fraction of a second when exposed to electrons from low-energy EB equipment [174,175]. In this process, multifunctional acrylates are used and have dual function: they accelerate the rate of polymerization and also cross-link the film. The method is suitable for packaging, coating, and ink applications [173].

5.4.2 Other Polymer Modifications

Other polymer modifications involve surface or bulk modifications and a majority of them are used in medical technology. Examples of these processes are:

- Modifications of surface of materials adapted for contact with human or animal tissues to impart biofunctional properties. These apply typically to ocular implants, surgical instruments, medical devices, or contact lenses [176].
- Enhancing the wear resistance of polymers (e.g., UHMWPE used for *in vivo* implants, such as artificial hip joints) [177].
- EB cross-linking of hydrogels based on synthetic PE oxide to produce materials for corneal prostheses [178].
- Producing PE microporous film with a porosity of 20–80% for battery separators [179].
- Three-dimensional processing of materials by EB to produce uniform isotropic irradiation of components, which are to be sterilized or bulk/surface modified [180].

An example of surface modification by EB irradiation was modification of PET surface by graft copolymerization with *N*-vinylpyrrolidone (NVP) in aqueous solution. The irradiation of PET films on the surface of NVP solutions was found to yield stable products with changed hydrophilic properties. The process was carried out at a commercially appropriate dose of 50 kGy providing an efficient surface hydroxylation with the EB energy in the range 500–900 keV. This energy afforded the accelerated electron penetration depth, sufficient for the modification of the opposite film side that was directly in contact with the NVP solution [181].

A comprehensive review of radiation techniques in the formulation of biomaterials was published by Kaetsu [182], covering the subjects of biocompatible polymers, immobilization of proteins, immobilization of cells, drug delivery systems, and future trends, including novel therapies, diagnosis methods, and immunological prevention methods.

References

[1] Bradley R. Radiation technology handbook. New York, NY: Marcel Dekker; 1984. p. 100.

[2] Charlesby A. Atomic radiation and polymers. Oxford, UK: Pergamon Press; 1960.
[3] Silverman J. In: Singh A, Silverman J, editors. Radiation processing of polymers. Munich: Hanser Publishers; 1992. p. 16.
[4] Patel GN, Keller A. J Polym Sci Phys Ed 1975;13:305.
[5] Chappas WJ, Silverman J. Radiat Phys Chem 1980;16:437.
[6] Liu Z, Markovic V, Silverman J. Radiat Phys Chem 1985;25:367.
[7] Luo Y, Wang G, Lu Y. Radiat Phys Chem 1985;25:359.
[8] Lyons BJ. Radiat Phys Chem 1986;28:149.
[9] Zoepfl FJ, Markovic V, Silverman J. J Polym Sci Chem Ed 1984;22:2017.
[10] Kashiwabara H, Seguchi T. In: Singh A, Silverman J, editors. Radiation processing of polymers [chapter 11]. Munich: Hanser Publishers; 1992.
[11] Barlow A, Biggs JW, Meeks LA. Radiat Phys Chem 1981;18:267.
[12] Fisher P. The short-time electric breakdown behavior of polyethylene, [annual report]. Washington, DC: National Academy of Science Publication; 1982.
[13] Awatami J, Tsunekawa Y. J Jpn Soc Mech Eng 1961;27(161):1113 [in Japanese].
[14] Oonish H, Takayama Y, Tsuji E. Radiat Phys Chem 1992;38:495.
[15] Salmon WA, Loan ADJ. Appl Polym Sci 1971;16:671.
[16] Oda E. Radiat Chem Phys 1981;18:241.
[17] Dadbin S, et al. J Appl Polym Sci 2002;86:1959.
[18] Makuuchi K, Cheng S. Radiation processing of polymers and its industrial applications. Hoboken, NJ: John Wiley & Sons; 2012. p. 122.
[19] Chappas WJ, Mier MA, Silverman J. Radiat Phys Chem 1982;20:323.
[20] Tomlinson JN, Kline DE. J Polym Sci 1967;11:1931.
[21] Arnold PM, Kraus G, Anderson Jr RH. Rubber Chem Technol 1961;34:263.
[22] McGinniss VD. In: Mark HF, Kroschwitz JI, editors. Encyclopedia of polymer science and engineering, vol. 4. New York, NY: John Wiley & Sons; 1986. p. 418.
[23] Sisman O, Bopp CD. Physical properties of irradiated plastics, ORNL-928. Springfield, VA: National Technical Information Service Operations; 1951.
[24] Baccaveda M, Fossini V. J Appl Polym Sci 1966;10:399.
[25] Wippler C. J Polym Sci 1958;24:585.
[26] Clough RL, Gillen KT. Radiat Phys Chem 1983;22:527.
[27] Posselt K. Kolloid Z Z Polym 1958;223:104.
[28] Miller AA. J Phys Chem 1959;63:1775.
[29] Davies K, Glover LC. Physical properties of polymers handbook. Woodbury, NY: American Institute of Physics; 1996. p. 568.

[30] Charlesby A. Atomic radiation and polymers. Cambridge, UK: Pergamon Press; 1962.
[31] Alexander P, Charlesby A, Ross M. Proc Royal Soc 1954;223:392.
[32] Lyons BJ. Radiat Phys Chem 1995;45:159.
[33] Tabata Y. Solid state reactions in radiation chemistry. In: Taniguchi conference. Sapporo, Japan; 1992. p. 118.
[34] Sun J, Zhang Y, Zhong X. Polymer 1994;35:2881.
[35] Oshima A, Ikeda S, Seguchi T, Tabata Y. Radiat Phys Chem 1995;45 (2):269.
[36] Lyons BJ. In: Scheirs J, editor. Modern fluoropolymers. Chichester, UK: John Wiley & Sons; 1997. p. 341.
[37] Korenev S. Radiat Phys Chem 2004;71:521.
[38] Florin RE. In: Wall LA, editor. Fluoropolymers. NY: Wiley–Interscience; 1972. p. 321.
[39] Bowers GH, Lovejoy ER. I&EC Product Res Develop 1962;1:89.
[40] Rosenberg Y, Siegmann A, Narkis M, Shkolnik S. J Appl Polym Sci 1992;44:783.
[41] Scheirs J. In: Scheirs J, editor. Modern fluoropolymers. Chichester, UK: John Wiley & Sons; 1997. p. 61.
[42] Carlson DP, West NE. U.S. patent 3738923; 1973.
[43] Suther JL, Laghari JR. J Mat Sci Lett 1991;10:786.
[44] Pae KD, Bhateja SK, Gilbert JR. J Polym Sci Part B Polym Phys 1987;25:717.
[45] Wang TT. Ferroelectrics 1982;41:213.
[46] Timmerman R, Greyson W. J Appl Polym Sci 1962;22:456.
[47] Yoshida Y, Florin RE, Wall LA. J Polym Sci Part A 1965;3:1685.
[48] Lyons BJ, Glover LC. Radiat Phys Chem 1990;35:139.
[49] Lyons BJ, Glover LC. Radiat Phys Chem 1991;37:93.
[50] McCune LK. Textile Res 1962;32:262.
[51] Weyers RE, Blankenhorn PR, Stover LR, Kline DE. J Appl Polym Sci 1978;22:2019.
[52] Hanks CL, Hamman DJ. Radiation effects design handbook, NASA-CR-1787, Sec. 3; 1971.
[53] Brown JR, O'Donnell JH. J Appl Polym Sci 1979;23:2763.
[54] Hegyazy EA, et al. Polymer 1992;33:2904.
[55] El-Naggar AM, Kim HC, Lopez LC, Wilkes GL. J Appl Polym Sci 1989;37:1655.
[56] Franta I. In: Franta I, editor. Elastomers and rubber compounding materials. Amsterdam: SNTL Prague and Elsevier; 1989. p. 25.
[57] Quirk RP, Morton M. In: Mark JE, Burak E, Eirich FR, editors. Science and technology of rubber, [chapter 11]. 2nd ed. San Diego, CA: Academic Press; 1994.

[58] Bhowmick A, Stephens HL, editors. Handbook of elastomers. 2nd ed. New York, NY: Marcel Dekker; 2001.
[59] Röthemeyer F, Sommer F. Kautschuktechnologie. Munich: Carl Hanser Verlag; 2001 [in German].
[60] White JL. Rubber processing. Munich: Hanser Publishers; 1995.
[61] Hamed G. In: Gent AN, editor. Engineering with rubber. Munich: Hanser Publishers; 1992. p. 20.
[62] Böhm GGA, Tveekrem JO. Rubber Chem Technol 1982;55(3):608.
[63] Flory PJ. Ind Eng Chem 1946;38:417.
[64] Böhm GGA, et al. J Appl Polym Sci 1977;21:3193.
[65] Smith TL. In: Eirich FR, editor. Rheology, theory and application, vol. 5 [chapter 11], New York, NY: Academic Press; 1969.
[66] Pearson DS. Radiat Phys Chem 1981;18(1−2):95.
[67] Lal J. Rubber Chem Technol 1970;43:664.
[68] Kuzminskii A, et al. In: Proc. international conference on peaceful uses of atomic energy. Geneva; 1958. p. 258.
[69] Kozlov VT, Klauzen NA, Tarasova ZN. Vysokomol Soedin 1968; A10(7):1949.
[70] Kaufmann R, Heusinger H. Makromol Chem 1976;177:871.
[71] Katzer H, Heusinger H. Makromol Chem 1973;163:195.
[72] Turner D. J Polym Sci 1958;27:503.
[73] Turner DT. J Polym Sci 1960;35:541 and Polymer (London) 1959;1:27.
[74] Mullins L, Turner DT. Nature 1959;183:1547.
[75] Mullins L, Turner DT. J Polym Sci 1960;43:35.
[76] Kozlov VT, Yevseyev AG, Zubov IP. Vysokomol Soedin 1969; A11(10):2230.
[77] Bauman RG. J Appl Polym Sci 1959;2:328.
[78] Roberts DS, Mandelkern L. J Am Chem Soc 1958;80:1289.
[79] Harmon DJ. Rubber World 1958;138:585.
[80] Harmon DJ. Rubber Age 1959;86:251.
[81] Dogadkin BA, Mladenov I, Tutorskii IA. Vysokomol Soedin 1960; 2:259.
[82] Nablo SV, Makuuchi K. Meeting of the rubber division of American Chemical Society. Pittsburg, PA; 1994 [paper #83].
[83] Chai CK, et al. Sains Malaysiana 2008;37(1):79−84.
[84] Blanchford J, Robertson RF. J Polym Sci 1965;3:1289 1303, 1313 and 1325.
[85] Petrov Ya, Karpov VL, Vsesoyuz T. Soveshchenia po radiatsionoi khim. Acad Sci USSR 1958;279.
[86] von Raven A, Heusinger H. J Polym Sci 1974;12:2255.
[87] Böhm GGA. In: Dole M, editor. The radiation chemistry of macromolecules, vol. II. New York, NY: Academic Press; 1972.
[88] Parkinson WW, Sears WC. Adv Chem Ser 1967;66:37.

[89] Golub MA. J Am Chem Soc 1958;80:1794.
[90] Golub MA. J Am Chem Soc 1960;82:5093.
[91] Zapp RL, Oswald AA. Rubber division meeting (Cleveland), Am. Chem. Soc. 1975 [paper #55].
[92] Tarasova ZN, et al. Kauch Rezina 1958;5:14.
[93] Harmon DJ. Rubber Age (London) 1958;84:469.
[94] Yoshida K, et al. J Macromol Sci 1980;A14:739.
[95] Okamoto H, Adachi S, Iwai T. J Macromol Sci 1977;A11:1949.
[96] Okamoto H, Adachi S, Iwai T. J Polym Sci 1979;17:1267 and 1279.
[97] Dogadkin BA, et al. Uses of radioactive isotopes and nuclear rad. in USSR. In: Proc. all-union conference. Riga; 1960.
[98] Jackson WW, Hale D. Rubber Age 1955;77:865.
[99] Pearson DS, Böhm GGA. Rubber Chem Technol 1972;45:193.
[100] Palalskii BK, et al. Vysokomol Soedin 1974;A16:2762.
[101] Alexander P, Charlesby A. Proc Royal Soc London 1955;A230:136.
[102] Alexander P, Black R, Charlesby A. Proc Royal Soc London 1955; A232:31.
[103] Berzkin BG, et al. Vysokomol Soedin 1967;9:2566.
[104] Turner DT. In: Bateman L, editor. The chemistry and physics of rubber like substances. New York, NY: John Wiley & Sons; 1963. p. 563.
[105] Karpov VL. Sessiya Akad. Nauk SSSR po Mirnomu Ispol. Moscow: Atom. Energii, Akad. of Sci; 1955.
[106] Slovokhotova NH, Vsesoyuz II. Seveshchania po Radiatsionoi. Moscow: Khim. Acad. Sci.; 1958. p. 263.
[107] Loan LD. J Polym Sci 1964;A2:2127.
[108] EXXON Chemical Brochure on Conjugated Diene Butyl Elastomer.
[109] Odian G, Bernstein BS. J Polym Sci Lett 1964;2:819.
[110] Geissler W, Zott M, Heusinger H. Makromol Chem 1978;179:697.
[111] Faucitano A, Martinotti F, Buttafava A. Eur Polym J 1969;12:467.
[112] Odian G, Lamparella D, Canamare J. J Polym Sci 1967;C16:3619.
[113] Geymer DD, Wagner CD. Polymer Prep Am Chem Soc Div Polym Chem 1968;9:235.
[114] Smetanina LB, et al. Vysokomol Soedin 1970;A12:2401.
[115] Pearson DS, Böhm GGA. Rubber Chem Technol 1968;45:193.
[116] Eldred RJ. Rubber Chem Technol 1974;47:924.
[117] Kammel G, Wiedenmann R. Siemens Forsch Entwicklungsber 1976; 5:157.
[118] El. Miligy AA, et al. Elastomerics 1979;111(12):28.
[119] Krishtal' IV, et al. Kauch Rezina 1976;(4):11.
[120] Tarasova NN, et al. Vysokomol Soedin 1972;A14:1782.
[121] Oganesyan RA. Kauch Rezina 1976;(1):34.
[122] Ito M, Okada S, Kuriyama I. J Mater Sci 1981;16:10.

[123] Polmanteer KE. In: Bhowmick AK, Stephens HL, editors. Handbook of elastomers. New York, NY: Marcel Dekker; 2001. p. 606.
[124] Ormerod MG, Charlesby A. Polymer 1963;4:459.
[125] Yasin T, Abmed S, Yoshii F, Makuuchi K. React Func Polym 2002;53:173.
[126] Bhowmick AK, Vijayabaskar V. In: 167th Technical meeting of the rubber division. San Antonio, TX: American Chemical Society; 2005 [paper #8].
[127] Charlesby A. Proc R Soc London Ser A 1955;230:120.
[128] Miller AA. J Am Chem Soc 1960;82:3319.
[129] Kilb RW. J Phys Chem 1959;63:1838.
[130] Delides CG, Shephard IW. Radiat Phys Chem 1977;10:379.
[131] Miller AA. J Am Chem Soc 1961;83:31.
[132] Przbyla RL. Rubber Chem Technol 1974;47:285.
[133] Mironov EI, et al. Kauch Rezina 1971;(6):19.
[134] Drobny JG. Technology of fluoropolymers. 2nd ed. Boca Raton, FL: CRC Publishers; 2009. p. 93.
[135] Arcella V, Ferro R. In: Scheirs J, editor. Modern fluoropolymers. Chichester, UK: John Wiley & Sons; 1997. p. 72.
[136] Tabata Y, Kojima G. Jpn Kokai JP 1973;73(38):465.
[137] Yamamoto T, Uchijima K, Ito Y. Jpn Kokai JP 1980;73(37):982.
[138] Ito M. J Jap Soc Rubber Ind 1986;59:169.
[139] Kaiser RJ, Miller GA, Thomas DA, Sperling LH. J Appl Polym Sci 1982;27(3):957.
[140] Vokal A, Pallanova M, Cernoch P, Klier I, Kopecky B. Radioisotopy 1988;29(5–6):426 CA 112(14):119976u.
[141] Logothetis AL. U.S. patent 5260351 to E.I. Du Pont de Nemours and Company; 1993.
[142] Caporiccio G, Mascia L. U.S. patent 5457158 to Dow Corning Corporation; 1995.
[143] Drobny JG. Handbook of thermoplastic elastomers. Norwich, NY: William Andrew Publishing; 2007. p. 150.
[144] Spenadel L, Grosser JH, Dwyer SM. US Patent 4,843,1·29 to EXXON Research and Engineering company (June 1989).
[145] Stine CR et al. US Patent 3,990,479 to Samuel Moore and Company (November 1976).
[146] Drobny JG. Rubber World 2005;232(4):27.
[147] Mehnert R, Bögl KW, Helle N, Schreiber GA. In: Elvers B, Hawkins S, Russey W, Schulz G, editors. Ullmann's encyclopedia of industrial chemistry, A22. Weinheim: VCH Verlagsgesselschaft; 1993.

[148] Mehnert R, Pincus A, Janorsky I, Stowe R, Berejka A. UV &EB curing technology & equipment. London and Chichester: SITA Technology Ltd./John Wiley & Sons; 1998. p. 22.
[149] Janke CJ, et al. U.S. patent 5877229 to Lockheed Martin Energy Systems; 1999.
[150] Shu JS, et al. U.S. patent 4657844 to Texas Instruments; 1987.
[151] Carroy AC, In: Conference proceedings of the 19th technology days in radiation curing process and systems, Le Mans, France; 1998.
[152] Heger A. Technologie der strahlenchemie von polymeren. Berlin: Akademie Verlag; 1990.
[153] Oldring PKD, editor. Chemistry and technology of U.V. and E.B. formulations for coatings, inks and paints. London: SITA Technology Ltd; 1991.
[154] Randell DR, editor. Radiation curing of polymers. London: The Royal Society of Chemistry; 1987.
[155] Radiation curing of polymeric materials. In: Hoyle CE, Kinstle JF, editors. ACS symposia series, vol. 417. 1990.
[156] Garratt PG. Strahlenhärtung. Hannover: Curt Vincentz Verlag; 1996, in German.
[157] Wellons JD, Stannett VT. J Polym Sci 1965;A3:847.
[158] Demint RJ, et al. Textile Res J 1962;32:918.
[159] Dilli S, Garrnett JL. J Appl Polym Sci 1967;11:859.
[160] Hebeish A, Guthrie JT. The chemistry and technology of cellulosic copolymers. Berlin: Springer Verlag; 1981.
[161] Garnett JL. Radiat Phys Chem 1979;14:79.
[162] Dworjanyn P, Garnett JL. Radiation grafting of monomers on plastics and fabrics. In: Singh A, Silverman J, editors. Radiation processing of polymers [chapter 6]. Munich Carl Hanser Verlag; 1992.
[163] Kabanov VY, Sidorova LP, Spitsyn VI. Eur Polym J 1974;1153.
[164] Huglin MB, Johnson BL. J Polym Sci 1969;A-1, 7:1379.
[165] Trommsdorf E, Kohle G, Lagally P. Makromol Chem 1948;1:169.
[166] Rengarajan LS, Parameswara VR. J Appl Polym Sci 1990;41:1891.
[167] Bhattacharya A, Misra BN. Progr Polym Sci 2004;29:767.
[168] Dilli S, Garnett JL. Aust J Chem 1970;23:1163.
[169] Ang CH, Garnett JL, Levot R, Long MA, Yen NT. In: Goldberg EP, Nakajima A, editors. Biomedical polymers, polymeric materials and pharmaceuticals for biomedical use. New York, NY: Academic Press; 1960.
[170] Stannett VT, Silverman J, Garnett JL. In: Allen G, editor. Comprehensive polymer science. New York, NY: Pergamon Press; 1989.
[171] Stannett VT. Radiat Phys Chem 1990;35:82.

[172] Garnett JL, Leeder JD. ACS Symp Ser 1977;49:197.
[173] Bett SJ, Garnett JL. SME technical paper FC-260. In: Proceedings of Radcure'87 Europe (Munich). Dearborn, MI: Society of Manufacturing Engineers; 1987.
[174] Simpson JT. Radiat Phys Chem 1985;25:483.
[175] Pacansky J, Waltman RJ. Prog Org Coat 1990;18:79.
[176] Goldberg EP, et al. U.S. patent 6387379 to University of Florida; 2002.
[177] Salovey R, et al. U.S. patent 6281264 to The Orthopedic Hospital, Los Angeles, CA and University of South California, Los Angeles; 2001.
[178] Perez E, Miller D, Merrill EW, U.S. patent 5,836,313 to Massachusetts Institute of Technology; 1998.
[179] Hasegawa T, Kondo T. U.S. patent 6127438 to Asahi Kasei Kogyo Kabushiki Kaisha; 2000.
[180] Nablo SV. U.S. patent 5825037 to Electron Processing Systems, Inc; 1998.
[181] Vasiljeva IV, Mjakin SV, Makarov AV, Krasovsky AN, Varlamov AV. Appl Surf Sci 2006;252(24):8768.
[182] Kaetsu I. In: Singh A, Silverman J, editors. Radiation processing of polymers [chapter 8 Radiation techniques in the formulation of synthetic biomaterials]. Munich: Carl Hanser Publishers; 1992.

Recommended Further Reading

Makuuchi K, Cheng S. Radiation processing of polymer materials and its industrial applications. Hoboken, NJ: John Wiley & Sons; 2012.

Industrial Radiation Processing with Electron Beams and X-rays. International Atomic Energy Agency. Vienna, Austria; 2011.

Drobny JG. Radiation technology for polymers. Boca Raton, FL: CRC Press; 2010.

Bhattacharya A, Rawlins JW, Ray P, editors. Polymer grafting and crosslinking. Hoboken, NJ: John Wiley & Sons, 2009.

Bhowmick AK, editor. Current topics in elastomers research. Boca Raton, FL: CRC Press; 2008.

Scheirs J, editor. Modern fluoropolymers. Chichester, UK: John Wiley & Sons; 1997.

Mark JE, editor. Properties of polymers handbook. Woodbury, NY: American Institute of Physics; 1997.

Clough RL, Shalaby SW, editors. Irradiation of polymers: fundamentals and technological applications [ACS symposium series 620], Washington, DC: American Chemical Society; 1996.

Garratt PG. Strahlenhärtung. Hannover: Curt R. Vincentz Verlag; 1996. p. 61 [in German].

Mehnert R. Radiation chemistry: radiation induced polymerization in Ullmann's encyclopedia of industrial chemistry, vol. A22. Weinheim: VCH; 1993.

Singh. A, Silverman J, editors. Radiation processing of polymers. Munich: Carl Hanser Verlag; 1992.

Bradley R. Radiation technology handbook. New York, NY: Marcel Dekker; 1984.

Böhm GGA, Tveekrem JO. The radiation chemistry of elastomers and its industrial applications. Rubber Rev 1982, Rubber Chem Technol 1982;55(3):575.

Charlesby A. Atomic radiation and polymers. New York, NY: Pergamon Press; 1960.

6 Industrial Applications of Ionizing Radiation

OUTLINE

Currently, there are more than 1400 high-energy electron beam (EB) accelerators in diverse industries around the world [1]. Large-scale industrial applications of ionizing radiation, mainly using EB irradiation equipment, started in the late 1950s, when Raychem introduced the production of PE heat-shrinkable tubing and W.R. Grace started to manufacture polyolefin packaging. At about the same time Goodyear and Firestone initiated investigation of modification of rubber compounds by EB irradiation for tire applications [2] and Ford Co. used EB irradiation for curing automotive coatings [3].

The chemical reactions of monomers, oligomers, and significant changes in properties of polymers induced by ionizing radiation discussed in previous chapters can be used for a variety of practical industrial applications. Besides the already-mentioned advantages in clean and safe technology, the almost instant conversion and excellent control of dosage and penetration depth EB

Drobny: Ionizing Radiation and Polymers.
DOI: http://dx.doi.org/10.1016/B978-1-4557-7881-2.00006-7

processing found its way into a variety of industrial applications such as wire and cable insulation, tire manufacturing, production of polymeric foams, heat-shrinkable films and tubings, curing of coatings, adhesives and composites, as well as in printing and other technological development. The equipment used must be matched to the given process. For example, for wire and cable, as well as for precuring of tire components and processing of rubber products, and polyolefin foams, where greater electron penetration is required, accelerators with energies ranging from 0.5 to 5 MeV and power rating less than 200 kW are used [4]. Coatings, adhesives, printing, and thin films can be processed by low-energy accelerators with energies typically in the 100–500 keV range. An additional dimension of ionizing radiation in industrial applications is the conversion of EB to X-rays. Some of the major industrial applications using EB processing of polymeric systems are given in Table 6.1. Figure 6.1 illustrates their relative volumes.

6.1 EB Process in Wire and Cable Technology

The wire and cable industry produces a large variety of conductors. For the most part, the materials used for the conduction of electricity are copper and aluminum. The primary function of polymeric materials, such as plastics, rubber, thermoplastic elastomers, and specialized coatings, is to provide electrical insulation. Other use, also very important, is as outer sheath (jacket) that protects the wire and cable insulation from effects of chemicals, moisture, ozone, and other environmental factors, and from abrasion and cutting. Specialized polymers and compounds improve flame resistance of the coated wires and cables.

The conductors produced by the wire and cable industry can be roughly divided into two basic groups: wires and cables. The difference between them is that a *wire* is a single conductor and a *cable* is a group of two or more insulated conductors. If there were not any insulation on the two conductors, then it would not be a cable; it would still be a single conductor which would be classified as a wire. There are four basic categories of wire and cable products. These categories consist of single conductor, multiconductor, twisted pairs, and coaxial cable. The most widely used conductor materials used are copper and aluminum.

There are four basic cable types: twisted pairs cable, multiconductor cable, coaxial cable, and fiber optics cable. *Twisted pairs cable* consists of pairs of conductors that are twisted together. This cable is specifically intended for signal carrying. Twisting the pairs of conductors gives the cable some immunity to interference. *Multiconductor cable* is cable that is made up of many insulated conductors (Figure 6.2). This type of cable is common in control applications but is almost never used in signal applications. *Coaxial cable* is

Table 6.1 Major Commercial Applications of the EB Processing of Polymeric Materials

Material/Substrate	Application	EB Process
Polyolefins and PVC, some elastomers	Wire and cable insulation	Cross-linking, 0.4–3 MeV, at approximately 10 kGy or higher
Elastomers	Tire manufacture, improved green strength of components and tire performance	Cross-linking with high-energy electrons
Polyolefins and PVC	Improving thermal stability, uniformity, and fine structure for packaging and insulation, rubber sheeting	Cross-linking with high-energy electrons
Wood impregnated with acrylic and methacrylic monomers	No-wear high-performance floors for high traffic areas	Polymerization by EB
Polyolefins	Heat-shrink films and tubing	Cross-linking
Polymeric films, metallic foils, paper, metal, wood	Curing of adhesives, coatings, and inks	Low-energy processing in the 100–500 keV range at 0.1–0.2 MGy
Polytetrafluoroethylene	Degradation into low molecular weight products ("micropowders") used as additives to coatings, lubricants and inks	High-energy irradiation at 200–400 kGy

the other popular configuration for cable. The signal on the two conductors in coaxial cable is not the same because the shield carries the ground and the signal. Since the signal is not the same on both conductors, this configuration is an unbalanced line. *Fiber optic cables* are divided into three kinds: plastic fiber, multimode fiber, and single-mode fiber. Plastic fiber is usually used in high-end audio signals. Multimode fiber is made of glass, ranges in diameters, and is used in data transmission. Single-mode fiber is so fine that it can be seen only under a microscope.

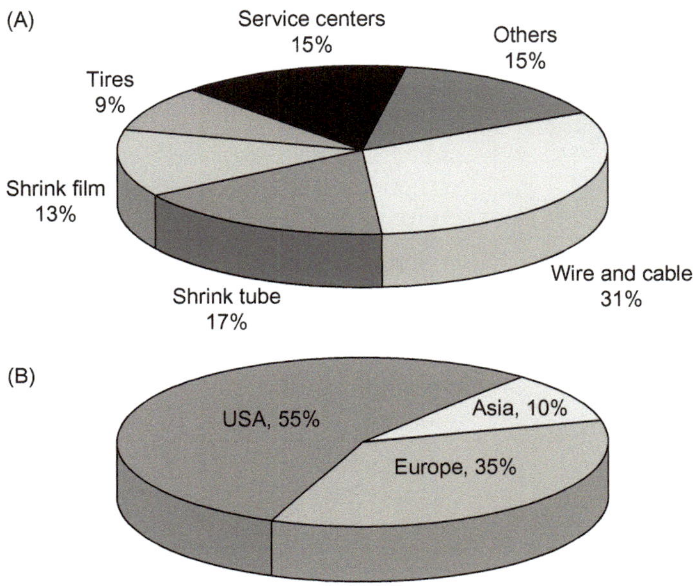

Figure 6.1 (A) World Electron Beam Market. (B) Electron Beam Market by Continent. *((A) Courtesy of A. J. Berejka.)*

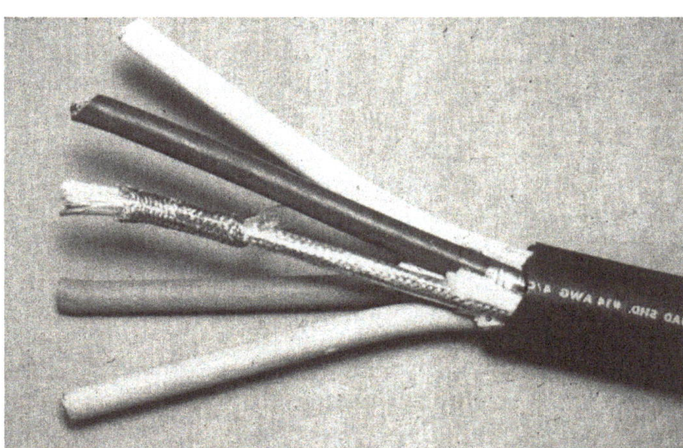

Figure 6.2 Example of a Multiconductor Cable. *(Courtesy of M.R. Cleland.)*

6.1.1 Equipment and Processes in Cable Manufacture

6.1.1.1 Extrusion

Crosshead extrusion process is widely used to coat wires and cables with a polymer-based insulation. The basic procedure includes pulling of the wire or cable to be coated at a uniform rate by a crosshead die (Figure 6.3), where it is covered with the molten plastic or hot rubber compound. This extrusion process for coating is used in most wires and cables that find usage in telecommunication and electrical applications along with the electronics industry. For more coatings, two extruders can also be used in tandem. Wire coating is generally done by the use of single-screw extruders, in which the crosshead extrusion process is carried out. The job of the extruder is to melt the resin and forward it to the die at an even and constant melt pressure and constant temperature. The crosshead extrusion process is carried out by using equipment in the line (Figure 6.4).

In this process, *primary insulation* is defined as the polymeric material applied directly onto the metal wire/cable to isolate the metal electrically. *Jacketing* (or *sheathing*) is referred to covering a wire or a group of wires with an insulating coating or jacket for nonelectrical protection. Jackets are usually put on to primary wires. The cooling of the extruded insulation or jacket is done in air or water.

Various polymers are used in wire coating applications by the crosshead extrusion process. The characteristics of these materials, which make them ideal for this purpose, are their flexibility, desirable electrical properties, ability to resist chemical, mechanical and environmental damage, and durability.

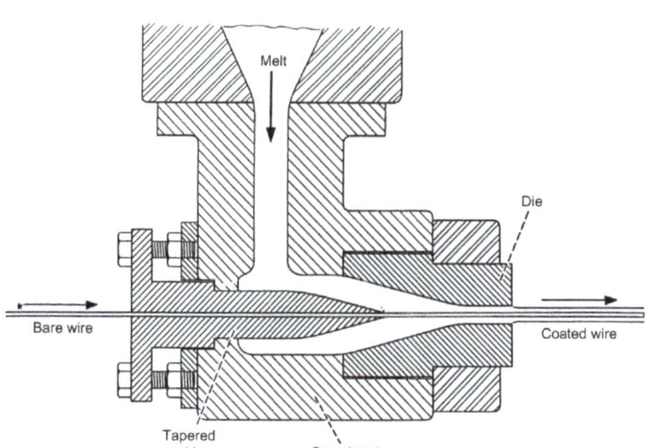

Figure 6.3 An Example of a Cross-head Die (Self-centering Design).

Figure 6.4 Wire and Cable Extrusion System with a Cross-head Die. *(Courtesy of Davis Standard LLC.)*

Rubber-based insulation has been used in W&C manufacture for nearly a century. Compounds for wire and cable are designed primarily to possess the required electrical properties for their intended service, and the ability to perform satisfactorily under their anticipated conditions of use. While physical properties are of secondary consideration, there are definite requirements imposed by a variety of specifications.

Jacket or sheath rubber compounds applied over insulation in cable structures provide the wear quality, resistance to elements, and overall strength required in service. Physical property requirements vary widely with the intended application, ranging from 1500 psi (10.3 MPa) tensile strength specified for flexible portable cords to 3500 psi (24.1 MPa) minimum for heavy-duty cables.

6.1.1.2 Vulcanization (Cross-linking)

The most widely used process, particularly in North America, to vulcanize (cross-link) rubber W&C compounds, is the *continuous vulcanization (CV operation)*. The production unit consists essentially of an extruder attached to a jacketed curing tube in which high-pressure steam is confined. The wire or cable with the compound applied by extrusion discharges directly from the extruder head into the steam tube through which it is conveyed under tension. Steam pressures of 1.4 MPa (200 psi) and higher are generally used and tube lengths may be 60 m (200 ft) or longer. The heated tube is often divided into several separately controlled heat zones, which can be adjusted up to temperature as high as 450°C (842°F). Compounds for continuous vulcanization are designed to cure in seconds at elevated temperatures and their handling and processing is somewhat more critical than in conventional rubber manufacturing.

The largest volume of polymeric materials used for wire and cable insulation are thermoplastics, namely PE and PVC [5] and to a lesser degree elastomeric compounds. The main reason of the prevalence of the PE and PVC in wire and cable insulation is their easy processing and relatively low cost. However, their main disadvantage is that their physical properties, such as plastic flow at elevated temperatures, environmental stress cracking, poor solvent resistance, and low softening temperatures [6], cannot always meet demands imposed on them by modern applications. Cross-linking of these materials improves their toughness, flexibility, and impact resistance, resistance to solvents and chemicals as well as increasing their service temperature [7,8].

6.1.1.3 Radiation Versus Chemical Cross-linking

There are essentially three cross-linking processes used in the wire and cable industry: two kinds of chemical cross-linking, employing organic peroxides or silanes respectively and radiation cross-linking by EB irradiation. Chemical cross-linking methods are still used for the improvement of PE and other polymer-based insulations for wire and cable [9,10]. However, the advantages of radiation cross-linking over the chemical processes have brought about its steady growth of about 10–15% annually [11] since the mid-1970s when high-energy and high-current electron accelerators became widely available for radiation processing [12,13]. Effects of these three cross-linking (curing) methods on various criteria applicable to wires and cables are given in Table 6.2. Today, more than 30% of the industrial EB processes in the world are used in the radiation cross-linking of wire and cable insulation [6].

The production line for continuous vulcanization of wires and cables by peroxide must be about 200 m (61 ft) long [5, p. 73]. A typical continuous wire and cable cure system is shown in Figure 6.5. The plant space required for radiation cross-linking using electron accelerators is considerably smaller. Moreover, energy consumption for the radiation process is much lower than that for the chemical cross-linking process [14]. The dose required for the desired degree of cross-linking is determined by the electron current of the accelerator and by the wire or cable speed; therefore, the process control is quite simple [15]. Depending on the source (most commonly steam) and the size of the product, regulation of the temperature of the vulcanization line, typically in the range 200–400°C (392–752°F), is more complicated than the control of the EB current. A modern EB curing line for wires and cables is shown in Figure 6.6. (Details about the equipment are given in Section 3.2.2.)

While chemical cross-linking is applicable only to PE and elastomers, EB radiation can be used for the cross-linking of PE, PE with flame retardants, PVC, some fluoropolymers, and some elastomers. The comparison of radiation and peroxide cross-linking in the wire and cable applications is given in Table 6.3.

The silane cross-linking of PE has been used mainly in low-voltage applications. It is a two-stage process and the curing time is dictated by the time required for the moisture to diffuse thoroughly into the insulation. Consequently, the process is rather slow. Another drawback of this system is the limited shelf life of the silane cross-linking agent and the significant cost of the compound [5, p. 74].

In general, conventional methods of cross-linking rubber, such as CV, involve chemical reactions between the elastomer and cross-linking agent often at very high temperature. On the other hand, irradiation by an EB is performed at ambient temperatures. Thus irradiation by ionizing beams offers a variety of advantages over the chemical curing method, such as:

- Considerable energy savings.
- Considerable space savings over the CV cure.
- The deleterious effects of steam, such as microvoids and bubbles, are all but eliminated.
- Irradiation is more versatile and offers more precise control over cross-link density; temperature- and moisture-sensitive materials can be processed.

Table 6.2 Effects of Curing Methods on Properties, Performance, and Relative Costs of Wires and Cables

Criterion	Radiation	Peroxide Cure	Silane Cure
Suitable size of wire or cable	Small	Large	Large
Compound cost	Low	Medium	High
Shelf life of compound	Long	Medium	Short
Production rate	Large	Small	Small
Degree of cross-linking	Medium	High	Low

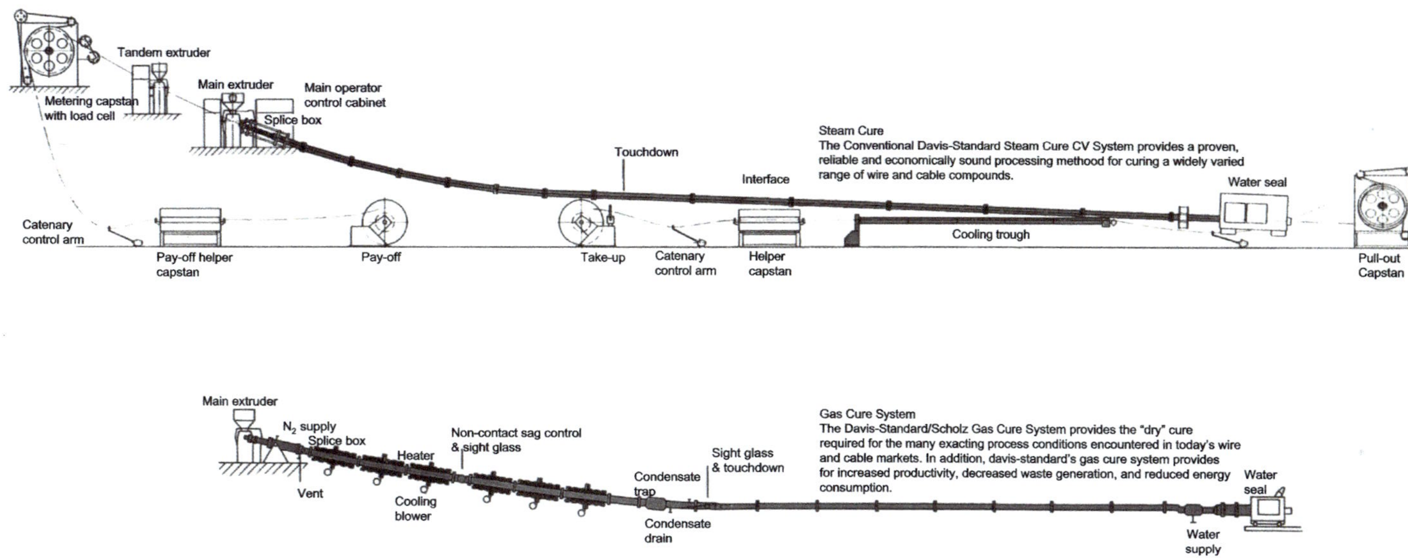

Figure 6.5 Wire and Cable CV Cure System. Top: Standard Steam Heating; Bottom: Gas Heating. *(Courtesy of Davis Standard LLC.)*

Figure 6.6 Modern EB Line for Wire and Cable Curing Easy-e-Beam™ System. *(Courtesy of IBA Industrial.)*

Table 6.3 Comparison of Radiation and Peroxide Cross-linking of Wire and Cable Insulation Materials

Property	Radiation Cross-linking	Peroxide Cross-linking
Energy consumption	Low	High
Line speed, m/min (ft/min)	Fast, up to 500 (1640)	Slow, max. 200 (655)
Processing	Off-line	Online
Factory space	Relatively small	Relatively large
Product sizes	Wide variety	Fixed
Process control	Current and line speed	Heat flow
Insulation materials[a]	LDPE, FRLDPE, HDPE, FRHDPE, PVC	LDPE
Maintenance cost	Low	High
Start-up	No scrap	100 m (330 ft) scrap at start
Voltage rating, kV	Up to 5	50
Capital cost	High	Moderate

[a]*FRLDPE, flame-retardant LDPE; HDPE, high-density PE; FRHDPE, flame-retardant HDPE; PVC, polyvinyl chloride.*

- Residual unreacted cross-linking agent in the chemical process will generally degrade electrical properties of the material.

Effects of cross-linking on dielectric constant values of selected polymeric insulating materials are given in Table 6.4.

Polymeric materials used for electrical insulation have to exhibit not only good electrical properties, but also good mechanical properties and good appearance, which are important. In addition to reliable electrical performance, thermal stability is important, mainly for miniaturized electronic and electrical equipment. Long-term stability at temperatures in the range between 90°C and 125°C (194–257°F) is required and short-term stability, determining the resistance to melted solder and cut-through resistance.

The most important performance tests on insulating materials are those for dielectric strength and insulation resistance. Wire and cable products are required to perform over different voltage ranges and under different conditions, such as a humid environment and after aging [16]. One of the tests is dielectric breakdown. Dielectric breakdown occurs when electrons detached from the molecule acquire sufficient energy in the electric field to yield secondary ionization and avalanche [17]. For most polymers, the dielectric strength can be as high as 1000 MV/m. However, under practical conditions, breakdown of polymer insulation occurs at much lower electric field strengths. For example, if the power dissipated in the insulating material raises its temperature enough to cause thermal stress, the dielectric breakdown occurs at much lower field strength. Another factor is surface contamination that can cause many polymeric insulators to break down by tracking on their surface [18].

Table 6.4 Effect of Cross-linking on Dielectric Constant Values of Selected Polymeric Insulating Materials

Insulating Material	Volume Resistivity (ohm cm)	Dielectric Constant	
		Uncured	Cross-linked
Low-density PE	10^{18}	2.3	2.0
High-density PE	10^{16}	NA	NA
Rigid PVC	10^{16}	NA	NA
Plasticized PVC	10^{11}–10^{14}	3.8	3.2
PP	10^{16}	2.3	NA
EPDM	10^{16}	3.1	3.6

Two other important electrical properties must be taken into consideration when polymers are used as insulation of high-voltage power cable [19] or electronic wires [20]. These properties are the dielectric constant and the dielectric loss factor, which characterize the energy dissipation in the insulation, the capacitance, the impedance, and the attenuation of electrical systems.

In contrast to metals and semiconductors, the valence electrons in polymers are localized in covalent bonds [21]. The small current that flows through polymers upon the application of an electric field arises mainly from structural defects and impurities. Additives such as fillers, antioxidants, plasticizers, processing aids, or flame retardants cause an increase of charge carriers, which results in a decrease of their volume resistivity [5, p. 76]. In radiation cross-linking, electrons may produce radiation defects in the material; the higher the absorbed dose, the greater the number of defects. As a result, the resistivity of a radiation cross-linked polymer may decrease [22].

Abrasion resistance and solder iron resistances are two other important properties required for a wire and cable insulator. These properties are critical for polymers used as sheathing or jacketing for hookup wire. Cross-linking improves both of them [5, p. 77].

High-voltage PE insulation thicker than 4 mm (0.16 in) tends to form voids due to heat accumulation, the evolution of hydrogen gas, and the discharge breakdown due to accumulation of excess charges [23] when irradiated. To eliminate this problem, the addition of prorads (such as multifunctional acrylate and methacrylate monomers) to reduce the required dose [24] is necessary. Some monomers, such as TAC and dipropargyl succinate [24], are effective in reducing the dose required for cross-linking, but they have an adverse effect on the dissipation factor of the insulation [25].

PVC, another widely used polymer for wire and cable insulation, cross-links under irradiation in an inert atmosphere; when irradiated in air, scission predominates [26,27]. To make cross-linking dominant, multifunctional monomers such as trifunctional acrylates and methacrylates must be added [5, p. 80] [28]. Comparison of standard and irradiated PVC wire insulation is given in Table 6.5.

Fluoropolymers such as copolymer of ETFE, PVDF, and polyvinyl fluoride (PVF) are widely used in wire and cable insulations. They are relatively easy to process, have excellent chemical and thermal resistance, but tend to creep, crack, and possess low mechanical stress at temperatures near their melting points. Radiation has been found to improve their mechanical properties and crack resistance [29].

Ethylene propylene rubber (EPR) has been also used for wire and cable insulation. When blended with thermoplastic polyolefins, such as LDPE, its processibility improves significantly. Typical addition of LDPE is 10%. EPR

Table 6.5 Comparison of Standard and Irradiated PVC Wire Insulations[a]

Property	Standard PVC, Not Irradiated	Irradiated PVC
Tensile strength, psi (MPa)	2000 (13.8)	4200 (29)
Elongation at break, %	200	170
Oxygen index	32	37
Solder iron, seconds to cut-through	<1	>600
Chisel cut-through, °C	85	108
Flexibility, psi (MPa)	23 (0.16)	80 (0.55)
Swelling in ASTM Oil #3	Good	Excellent

[a]*Insulated wire with 10 mil (0.25 mm) wall of insulation.*

and terpolymers with high PE content can be cross-linked by irradiation [5, p. 80]. An example of an EPDM-based EB-curable fire-resistant formulation for wire insulation is given in Table 6.6.

6.1.2 Radiation Equipment and Process in W&C Manufacture

Radiation cross-linking of wire and cable insulation requires, in general, medium-energy electrons in the range 0.5–2.5 MeV. The electrostatic EB systems using high DC voltage to accelerate electrons have been found to be most suitable [5, p. 81]. Cross-linking of wire and cable insulation results from the dissipation of the energy carried by fast electrons penetrating into the insulation. As pointed out in Section 2.2, the depth of penetration is governed by the accelerating voltage. For that reason, the equipment used has to have sufficient beam energy, since it defines the maximum thickness of the insulation, which can be penetrated and effectively cross-linked by the electrons [31,32]. The effective depth (optimum penetration) of the EB is defined as the thickness in which the exit plane dose equals the incident surface dose. The ratio of the peak dose to the entrance dose is defined as dose uniformity [5]. Because of the nature of the process [33], the energy imparted to the insulating material is not uniformly distributed. However, the physical properties of the insulation modified by radiation cross-linking are not very sensitive to the variation of the dose in depth. Thus the heterogeneity of the dose should not be of great concern and the maximum-to-minimum dose ratio of 1.5 or less is usually satisfactory [5]. The EB curing system shown in Figure 6.6 is used typically for thin wall wires from 24 to 9 G (0.22–6 mm^2 cross-sectional

Table 6.6 Example of an EB-Curable Fire-Resistant Insulating Material

Material	Amount (phr)[a]
EPDM (with ethylene norbornene, diene cure site)	100
Hydrated aluminum oxide (filler, flame retardant)	100
Paraffinic process oil (plasticizer)	50
Silane A172 (coupling agent)	2
Antioxidant	1
TMPT (75%) (prorad)	3
Total	256

[a]*phr, parts per hundred parts of rubber (commonly used in rubber compounding).*
Source: *Ref. [30].*

area. Maximum system processing speed: 1000 m/min (3200 ft/min), scan width: 0.91 m (36 in) (more details in Section 3.2.2).

The volume resistivity of PE is very high and the electrons that come to rest in the insulation layer cannot be easily removed. Therefore, it is necessary for the electrons irradiating the wire or cable to penetrate through the insulation to ensure that no charge is accumulated in the wire or cable, causing dielectric breakdown [34]. PVC has a lower resistivity than PE; therefore, charges during the irradiation can leak from the insulation through the wire or cable conductor to ground. The penetration depth (or range of electrons) for a given EB depends on the thickness of the wire or cable insulation and the diameter of the conductor can be calculated [5, p. 82] [35]. Usually, a penetration of twice the radial thickness of the insulation is sufficient [5, p. 84].

Irradiation cross-linking of small wires and cables with insulation thickness around 1 mm (0.040 in) is done under a scanner. The wire or cable is strung between two drums to form a figure of eight (Figure 6.7) and is irradiated on both sides. The dose is controlled automatically by a servolink.

Cables with large conductor cross section (e.g., 150 mm^2) and thicker insulation (2 mm or 0.080 in) require a multiside irradiation [5, pp. 84,35,36]. A schematic of four-sided irradiation using a single accelerator is shown in Figure 6.8 [5, p. 84], where the scanning horn is placed over four wire/cable strands twisted between rollers. The actual production fixture for a multipass irradiation is shown in Figure 6.9. The cable speed, the rotation speed of the twisting cable, and the scanner frequency are coordinated by a servomechanism [5, p. 84]. Some high-frequency accelerators can provide deflected EBs capable of irradiating the cable from three directions. In such an arrangement, the required electron penetration equals the required thickness of insulation; consequently, electron accelerators of lower energy can be used [5, p. 84].

The doses required for EB cross-linking of PE and PVC (compounded with prorads) are 100–300 and 40 kGy, respectively. The number of passes of wire or cable through the beam has its limitations, since sufficient spacing is required between passes to avoid overlap or shadows. The overall energy utilization coefficient can vary anywhere between 0.20 and 0.80 [5, p. 81]. Leading manufacturers of EB equipment provide formulas for the calculation of required energy [37] and processing rate for the given dimensions of a wire and cable [37–41].

6.1.3 Materials

Depending upon the end-use requirement, wire jacketing is most often made from formulated PE. Blends of PE and ethylene–propylene rubber are used if greater flexibility is needed, especially as the diameter of the jacketing increases as in the case with cables. Another possibility is EPDM rubber

Figure 6.7 Two-Sided Irradiation of a Wire or Cable, the "Figure Eight" Configuration. *(Courtesy of M.R. Cleland.)*

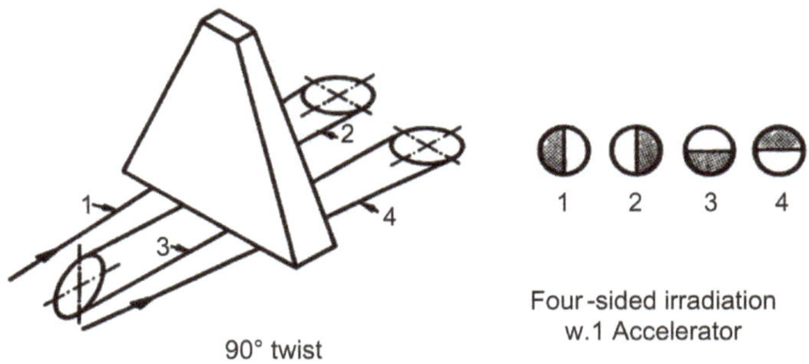

Figure 6.8 Schematic of the Four-Side Irradiation of W&C.

Figure 6.9 Multipass Wire Irradiation Fixture for Wire or Small Tubing. *(Courtesy of IBA Industrial.)*

with ENB monomer. Table 6.6 presents a typical radiation cross-linkable jacketing formulation [30].

Hydrated aluminum oxide is a preferred flame retardant that liberates its water of hydration when exposed to flames in contrast to chlorinated materials which give off toxic gases as by-products. The paraffin oil is a processing aid that enhances the ability to extrude such materials. The silane is a coupling agent that improves the interaction between the polymers and the aluminum trihydrate. TMPTMA enhances the radiation response.

When enhanced temperature resistance is required, PVDF or other fluoropolymers are used [42]. Fluoropolymers have the advantage of being oil resistant and flame retardant, but are also more expensive base materials. PVDF is one of the materials that are very easy to cross-link by EB radiation.

6.1.4 Recent Developments and Trends

Radiation cross-linking of wire and cable insulations has been growing steadily over the past 33 years. On a weight basis, more wire insulation than cable insulation has been cross-linked by EBs, most of the wires being used in electronics. Recently most growth has been in flame-retardant insulation. Besides PE and PVC, chlorinated PE and chlorosulfonated PE (Hypalon®) have been used [42]. PP, similar to PE in price, but stronger, can also be cross-linked by EBs but requires prorads, such as TMPTA. These reduce the dose required for the cross-linking of PP considerably [43]. Another unique radiation cross-linked insulation is polymeric foam wire and cable insulation [5].

6.2 EB Process in Tire Technology

A pneumatic tire is composed of several components (Figure 6.10), which must retain their shape and dimensions during the process of being built and during vulcanization in the mold. Such components are innerliner, body plies and tread ply skims, belts, beads, and chafer strips. When the various parts of a tire are partially cured by radiation, they will not thin out or become displaced during the construction of the tire, nor will they thin out and flow during the vulcanization in the mold as would components that were not irradiated. In some cases, material reduction is possible and more synthetic

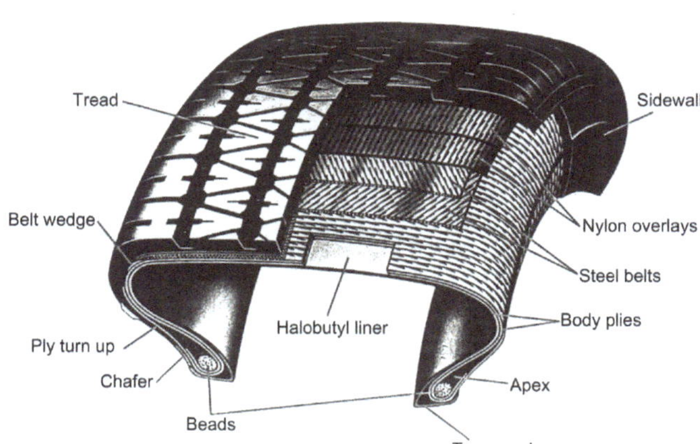

Figure 6.10 Cross Section and Components of a Radial Tire.

rubber can be used in place for the more expensive NR without a loss of strength [44].

Body plies of a tire are various rubber-coated cord fabrics. These are subjected to stresses when the tire is built. If a body ply does not have sufficient green (i.e., uncured) strength, the cords will push through the rubber coating (skim) during the building and vulcanization of the tire resulting in irregular cord placement or a defect in the tire. For example, damage can occur from turning the ply around the bead during the building of the tire when the skim compound does not have adequate green strength. Green strength may be defined as that level of cohesive strength which allows an essentially uncross-linked polymer-based composition to deform uniformly, under stress, without sagging, or nonuniform thinning (necking).

Tires are built most commonly on expandable drums. The construction of a tubeless tire starts with placing the innerliner on the building drum. Then the different plies and other components are added including the tread, which is applied last. When the drum is expanded to shape, the green (i.e., uncured) tire, the cord spacing in the body plies, must be maintained. In tread plies, correct cord placement must be maintained during the change in their angle occurring during the shaping in the vulcanization process. Thus the tread ply skim compound, like the body skim compound, has to have adequate green strength. Another important property of the rubber material in the tire structure is the building tack, which is essentially adhesion to itself (autohesion). The level of building tack is reduced with the degree of cross-linking of the rubber compound. Thus, the green strength and building tack must be in a good balance.

Then the tire is generally expanded into toroidal shape by air pressure. Additional expansion occurs in the heated curing mold during vulcanization of the tire. During the expansion of the green tire and during vulcanization, the innerliner tends to decrease in thickness (thins out) and to flow considerably, especially in the shoulder area, depending on the green strength of the innerliner compound. Since the innerliner has the function of maintaining the inflation pressure of a tire during its service, its gauge is critical to the air permeation. Thus, maintaining its thickness in the curing process is very important. Normally, this is assured by using a thicker innerliner than necessary. Partial cross-linking of the compound by irradiation makes this unnecessary, resulting in saving material and cost.

The degree of partial cure has to be precisely controlled, as pointed out earlier, in order to assure sufficient green strength and sufficient building tack necessary for the construction of the tire. The EB process is ideally suited for that. The partial cure or precure is done by simply placing the tire components on a conveyor belt, which is passed under the EB source, and exposing them to the proper irradiation dose. The dose depends on the particular elastomer (or elastomer blends), type and amount of fillers, oils, antioxidants, and other compounding ingredients that more or less affect the radiation response of the compound. Typically, a dose of 0.1–20.0 Mrad (1–200 kGy) is required [44]. A cross section of a tire body ply with controlled depth of partial cross-linking is shown in Figure 6.11.

The effects of EB irradiation on the rubber compounds for tires can be readily evaluated by measuring green strength and recovery. Green strength is measured in a similar fashion to tensile strength: straight-sided specimens typically 12.5 mm (0.5 in) wide and 2.5 mm (0.1 in) thick are stretched by a tensile tester (e.g., Instron) and the peak stress or tensile strength is recorded. An example of the effect of irradiation on green strength is shown in Figure 6.12. The laboratory test results from an uncured sheet can show

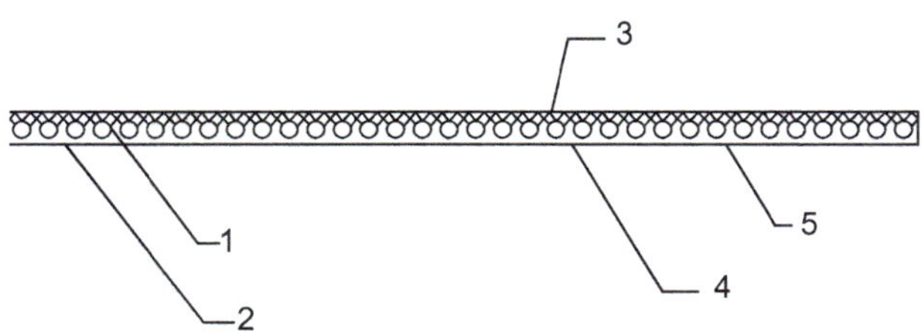

Figure 6.11 Cross Section of a Tire Body Ply with Controlled Depth of Partial Cross-linking. 1, Reinforcing Cord Fabric; 2, Skim Coat; 3, Partially Cross-Linked Skim Coat; 4, Building Tack Preserved; 5, Bleed Cord.

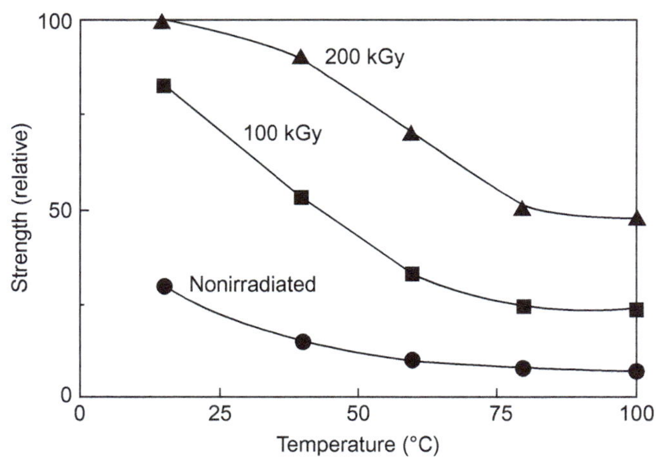

Figure 6.12 Effect of Irradiation Dose on Green Strength at Increasing Temperature [45, p. 159]. *(Reprinted with permission.)*

whether the material has sufficient green strength for the given tire component if the slope of its tensile stress–strain curve remains positive beyond the maximum extension to which the component will be subjected before it is vulcanized [44]. For example, the green strength of the inner liner compound for a radial tire can be shown to be sufficient if the slope of its tensile stress–strain curve is positive throughout the 100–300% elongation range, because maximum extension of the inner liner during the forming of the tire falls into the lower end of this range. If extension in the crown section of the tire were to exceed the uniform deformation capability of the liner, nonuniform thinning would occur [46].

The recovery may be evaluated by the conventional Williams plasticity test (ASTM D926), Procedure 2.2.1 at 100°C (212°F). Other tests may be used, although the ones mentioned here were found to correlate best with the behavior of the rubber compound and tire components during the building and vulcanization of the tire. Results from green strength and recovery tests from two different compounds reported in Ref. [44] are given in Table 6.7 and green strength of irradiated ply skim stocks with different compositions is given in Table 6.8. Physical properties from vulcanized bromobutyl inner liner compound (unaged and aged) are given in Table 6.9.

The irradiated components can be placed adjacent to sulfur-containing unvulcanized elastomeric compounds, and when necessary, coated with suitable adhesive between the contacting surfaces to assure good bond after final cure of the tire [44].

The equipment and technology have developed significantly over the past two decades during which the high-energy EB process has been in use.

Table 6.7 Effect of Irradiation on Green Strength and Recovery of Two Different Tire Compounds

		Dose (kGy)				
Compound	**Property**	**0**	**50**	**100**	**150**	**200**
Innerliner	Green strength, kg[a]	1.26	6.94	9.07	1.39	15.33
	Recovery, %[b]	8.0	31.5	42.5	80.0	–
Chafer strip	Green strength, kg[a]	10.34	15.92	19.10	23.59	26.40
	Recovery, %[b]	9.0	39.0	48.0	54.0	–

[a]*Peak value in tensile test.*
[b]*Plasticity. Williams (ASTM D928, Procedure 2.2.1).*

Table 6.8 Green Strength of Irradiated Ply Skim Stocks with Different Elastomer Compositions (kg)[a]

Type of Skim Stock	Nominal Radiation Dose (kGy)				
	0	10	20	30	50
NR Solution SBR 50/50	1.45	2.09	2.99	3.86	5.12
100% Solution SBR	0.27	0.36	0.77	1.45	–
SBR/BR 75/25	0.68	–	–	–	11.66
100% BR	1.59	3.86	6.17	8.26	–
100% SBR	0.18	0.97	2.63	4.26	–

[a]*From peak values of tensile strength test.*

Table 6.9 Physical Properties from Vulcanized Bromobutyl Inner Liner Compound

(Unaged)					
Property	Nominal Radiation Dose (kGy)				
	0	10	20	30	50
Hardness, Durometer A	53	50	48	46	46
100% modulus, MPa	1.2	1.1	1.0	0.8	0.9
300% modulus, MPa	4.8	4.3	4.3	3.6	4.0
Tensile strength, MPa	9.9	8.7	8.0	6.9	7.5
Elongation at break, %	640	600	610	670	620
Peel adhesion at RT, kN/m	4.2	3.2	6.4	3.8	3.6
Peel adhesion at 100°C, kN/m	6.3	3.7	3.9	1.1	1.4
Fatigue, kcy (cured to Rheometer optimum)	130	165	160	155	110
(Aged 240 h at 125°C)					
Hardness, durometer A	60	53	55	54	55
100% modulus, MPa	2.2	1.8	1.8	1.8	1.9
300% modulus, MPa	6.9	6.4	6.4	6.3	6.4
Tensile strength, MPa	8.4	7.5	7.5	7.2	7.3
Elongation at break, %	420	390	400	390	380
Fatigue, kCy (cured to rheometer optimum)	25	60	50	45	40

The original equipment was shielded by concrete vaults, which were 70–80 ft long, 35–40 ft high, and 16–20 ft wide. The walls of the treatment area might have been 4–6 ft thick. Pay-off and take-up equipment would add another 60–80 ft to overall length of the system. The tire parts were treated almost exclusively off-line [5,44,46]. Due to a strong push by tire manufacturers to make the equipment more user-friendly, the accelerators were redesigned to produce radiation in the forward direction, which required radiation shielding only in the front and on the sides [44]. Furthermore, the vault was placed below the ground level and consequently the factory floor provided the bulk of the shielding. Further improvements were in designing the shielding structure from a combination of lead and steel replacing the traditional concrete vault. Currently, self-shielding low- to medium-energy (500–800 keV) EB machines are used to partially cross-link the body plies or innerliner.

The components that are supported by cords or fabric can be treated online easily by using a system of pulleys. This, unfortunately, is not yet possible for unsupported components, such as innerliner and other low-tension materials [47,48].

The current technology is very cost effective. If applied correctly in-line with other processes, such as extrusion or calandering, it can produce substantial savings by the reduction in material weight or the weights of the tire components. The reduction of overall thickness can be as much as 20% [48]. A saving of $0.29 was reported by irradiation of tire components owing to the gauge reduction and a decrease of the ratio of natural to synthetic rubber [49].

6.3 EB Process in the Manufacture of Polyolefin Foams

The thin, highly stretched expanding foam cells of thermoplastic melts are unstable and are bound to rupture at elevated temperatures when not stabilized. This is particularly important when chemical blowing agents are used, since they require relatively high temperatures to decompose. One of the methods of stabilization of polyolefin melt is cross-linking. Cross-linking not only stabilizes bubbles during expansion but also enhances protection of the cellular product from thermal collapse, which is necessary in some applications.

EB irradiation is one of the methods of cross-linking in this process [50]. The other methods use peroxide [51,52], multifunctional azide [53–56], or an organo-functional silane [57]. PE resins respond to EB irradiation well since the rate of cross-linking exceeds significantly the chain scission. PP is prone to β-cleavage, which makes it difficult to cross-link by a free-radical process [57]. For that reason, PP resins require cross-link promoters, such as vinyl monomers, divinyl benzene [58,59], acrylates and methacrylates of polyols [60,61], polybutadiene, and others. These compounds promote cross-linking of a polyolefin polymer through inhibiting the decomposition of polymer radicals [62], grafting [63], and addition polymerization-type reactions [57].

Optimum cross-linking is the most critical requirement for optimum foam expansion. The *gel level* or *gel fraction* is defined as the fraction not soluble in boiling xylene [64]. Excessive cross-linking restricts foam expansion, while insufficient cross-linking results in bubble rupture [65]. The window of optimum cross-linking is fairly narrow; the gel level at the inception of gel expansion needs to be about 20–40%, preferably 30–40% [66,67]. An LDPE resin with a lower value of melt index (higher molecular weight) and a lower density becomes more efficiently cross-linked. A lower density LDPE has more chain branching and therefore more tertiary hydrogens, which provide cross-linking sites. Other factors influencing cross-linking are molecular weight distribution and long-chain branching [66–68]. For typical PE, a dosage ranging from 10 to 50 kGy is sufficient to attain the final gel level of 30–40%.

The selection of EB equipment depends on the foam thickness and production rate. Foam manufacturers throughout the world are using equipment with electron accelerators with voltages ranging from 0.5 to 4 MV and power ratings 10–50 kW. The penetration depth of a 1 MV unit is approximately 3 mm (0.12 in). Irradiation on both sides doubles the thickness capability [69].

6.3.1 Foam Expansion and its Control

The cross-linked sheet is expanded by means of a chemical blowing agent such as azodicarbonamide, which decomposes in the temperature range between 200°C and 210°C (392–410°F). In the radiation cross-linking process [70,71], the cross-linking reaction is completed before the decomposition of the blowing agent and foam expansion. Since the decomposition of the blowing agent and foam expansion occur simultaneously, this process can run twice as fast as peroxide cure, since the heating rate is not restricted by the cross-linking rate [66]. The cell size is easier to control in the cross-linked sheet since the cross-links provide the restricting force against cell growth. Thus, radiation cross-linking favors cell nucleation and fine uniform cells (typically 0.2–0.3 mm in size) are easily achieved [72].

6.3.2 Manufacturing Processes

Cross-linked polyolefin foam sheet is produced by two methods using chemical cross-linking and by two methods of radiation cross-linking. The two well-established manufacturing processes for polyolefin foams using radiation cross-linking are the *Sekisui* process and the *Toray* process. The differences between these two manufacturing methods are mainly in the expansion step, which is almost always done separately. However, the production of the foamable sheet and the cross-linking step are essentially similar. The first step is a uniform mixing of the blowing agent into the polymer melt, followed by sheet extrusion in an extrusion line. Then the cross-linking to the desired degree follows. The flow diagram for this process is shown in Figure 6.13.

The Sekisui process employs a vertical hot air oven. The foamable sheet is first preheated in the preheating chamber by IR heaters to about 150°C (302°F) while being supported by an endless belt. The preheated sheet is then expanded in the foaming chamber to temperatures exceeding 200°C (392°F). The expanding sheet supports itself by gravity in the vertical direction. A special tentering device is used to prevent the development of wrinkles in the transverse direction [73]. The advantages of the Sekisui process are the capability of manufacturing a thin sheet and the low-energy consumption inherent in the vertical oven [72, p. 231]. A schematic of the Sekisui foaming oven is shown in Figure 6.14.

In the Toray process, the foamable sheet is expanded as it floats on the surface of molten salts and is heated from the top by IR lamps. The molten salt mixture consists of potassium nitrate, sodium nitrate, and sodium nitrite [70]. The salt residues from the surface of the foam are blown off by hot air and stripped in water. The Toray process is suitable to produce cross-linked PP foam sheet as well as PE foam sheet. In fact, Toray was the first company to produce commercial PP foam [74]. A schematic of the Toray process is shown in Figure 6.15.

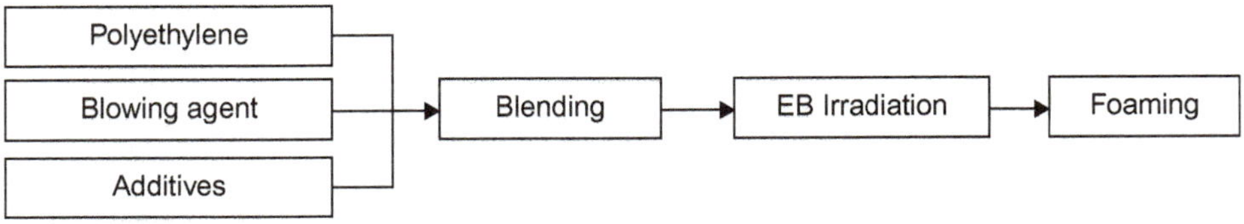

Figure 6.13 Flow Diagram of the Manufacturing Process of Polyolefin Foam Using Radiation Cross-linking.

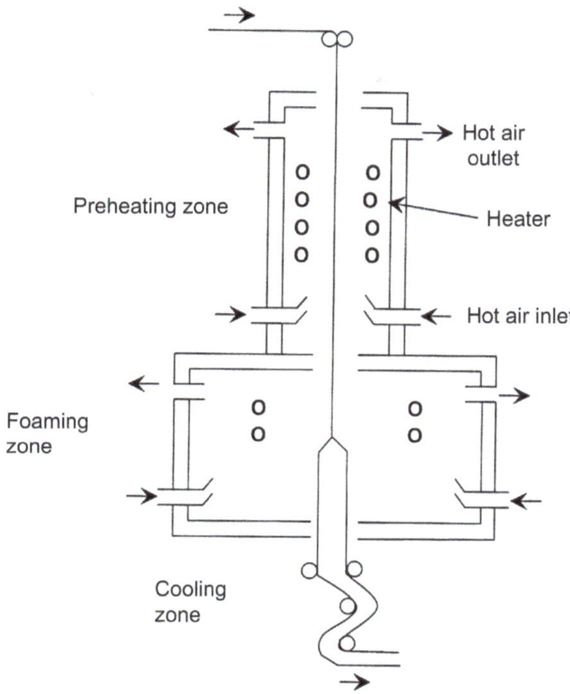

Figure 6.14 Schematic of Sekisui Foaming Oven for Polyolefin Foams.

6.3.3 Comparison of Chemical and Radiation Processes

Both processes have advantages and disadvantages. Currently, they have about equal share of the global market. The initial investment cost in the irradiation equipment is high, but it is considerably more productive than chemical cross-linking, having the additional advantage of uniform product quality and flexibility of feedstock selection. The product from radiation cross-linked polyolefins is thin with fine cells and smooth white surfaces. A comparison of chemical and radiation processes is given in Table 6.10 [66,72, p. 234].

6.3.4 Applications of Polyolefin Foams

Correctly foamed polyolefins consist of small regular closed cells. Because of this they have a high heat and sound insulation capacity and excellent shock absorbing capability. They are used for the insulation of

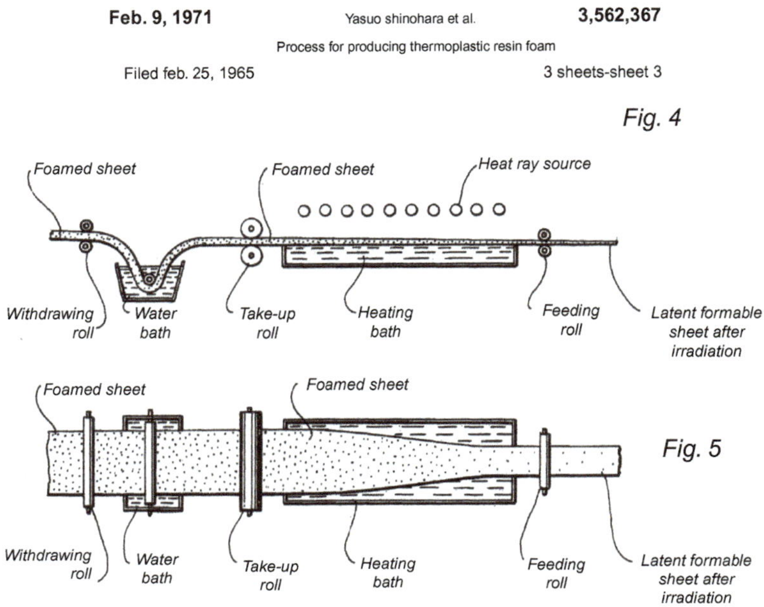

Figure 6.15 Schematic of Toray Process for the Production of Polyolefin Foams [69]. Top: Side view, Bottom: View from the top.

central heating pipes, in the packaging industry, and in sports and leisure articles for protection of the head, knee, shin, and elbow. In the health-care industry, polyolefin foams are used as backing for medical devices.

In automotive applications, polyolefin foams are used for safety and protection in dashboards and door panels (Figure 6.16), and most notably as cushioning under the interior header.

6.4 EB Process in the Production of Heat-Shrinkable Materials

Polyolefins, especially PE, can be cross-linked into a material, which is elastic when heated. The structure of polyolefins, normally entangled long chains, includes crystalline and amorphous regions. Upon heating above the crystalline melting point of the polymer, the crystalline regions disappear.

When cross-linking the polymer, a three-dimensional network is formed. After heating the cross-linked material above its crystalline melting point, the elastic network can be stretched. If the material is cooled in the stretched

Table 6.10 Comparison of Radiation and Chemical Cross-linking Processes for Producing Polyolefin Foams

Item	Radiation Cross-linking	Chemical Cross-linking
Process control	Easy	Difficult
Production rate	High	Slow
Equipment	Cumbersome	Simple
Cost	Decreases with production volume	Relatively constant
Blowing agent selection	Easier	More difficult
Product thickness, mm	3–6	5–16
Typical cell size, mm	0.2–0.4	0.5–0.8
Cross-link level, %	30–40	60–70

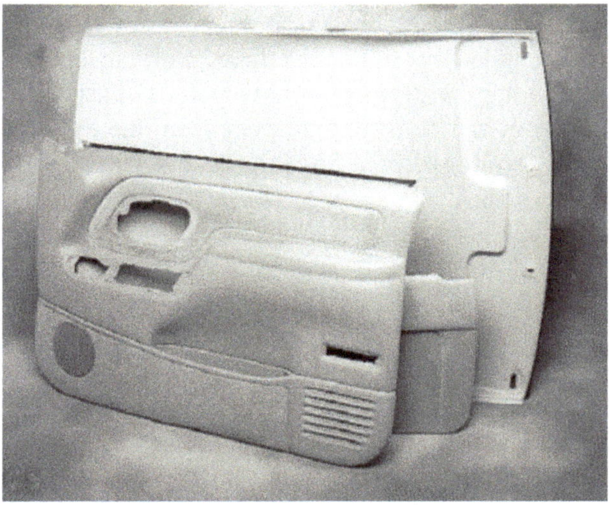

Figure 6.16 Cross-Linking PE Closed Cell in Automotive Application *(Courtesy of A.J. Berejka.)*

state below its crystalline melting point, the crystalline regions reform and the material retains its deformation. If it is heated again above the melting temperature, it will shrink back to its original state. This phenomenon is referred to as the memory effect and is used for heat-shrink tubing and in packaging.

6.4.1 Heat-Shrinkable Tubing

Heat-shrinkable tubing is made typically from polyolefins, PVC, polyvinyl fluoride, PTFE, their blends or blends with other plastics, and elastomers. The formulations may be designed for chemical resistance, heat resistance, flame resistance, etc. [75].

There are essentially two methods to produce shrinkable tubing from an extruded tube. One of them involves expanding the tube, which had been cross-linked by radiation and heated to temperatures above its crystalline melting point into a sizing tool. The diameter of the sizing tool determines the amount of stretch. The stretched tubing is then cooled down to temperatures below the crystalline melting point in the sizing tool thus "freezing-in" the strain. This method is described in the original Raychem patent (Figure 6.17). The patent specifies the irradiation dose to be at least 2×10^6 rad (20 kGy) [76]. Current practice is to irradiate PE tubing with electrons at 1.0–2.0 MeV and at a dose of 200–300 kGy. Excessive doses should not be applied, otherwise the material will have a poorer aging resistance [77].

The second method is designed to expand the tubing heated above its crystalline melting point into a stationary-forming tube or mold by air pressure. The tubing is surrounded by a moving tape over the entire length of the forming device. The tape is formed from a material that does not adhere to the cooled plastic so it can be readily stripped from the expanded tubing as it emerges from the forming device. An alternative patented process [78] is described in Figure 6.18.

The plastic memory effect of cross-linked heat-shrinkable tubing is demonstrated in Figure 6.19. Figure 6.20 depicts the process of producing heat-shrinkable PE tubing and the use of the tubing to cover a wire joint. In Step I, the tubing is extruded and irradiated to obtain a gel fraction of 40%. In Step II, the irradiated tubing is heated up to 140°C (285°F) and expanded by the use of vacuum or pressure to about twice its original diameter, holding its length constant. Steps III and IV show the use of heat-shrinkable tubing to cover a joint of two wires. The tubing is centered over the joint and heated by a hot air gun to a temperature above the crystalline melting point of PE. The sleeve shrinks and becomes a form-fitting cover for the joint [79].

Thin walled (approximately 0.010–0.040 in, or 0.25–1.0 mm) tubing is usually selected by continuous service temperature, which may range from

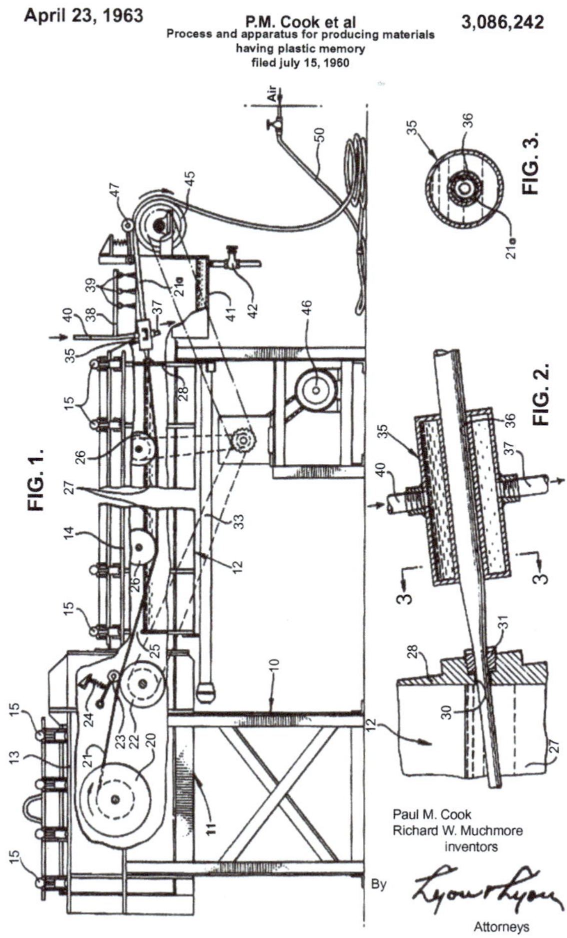

Figure 6.17 Original Raychem Patent for Heat-Shrinkable Tubing [76].

90°C (194°F) for some PVC tubing up to 250°C (482°F) for tubing made from PTFE. The main use of thin-shrink tubing is for electrical and electronic applications.

Tubing with thicker walls (typically in the range 0.080–0.170 in, or 2–4 mm) is fabricated mainly from polyolefins and is used to cover splices in telecom, CATV, and electric power industries. Often such tubing is combined with mastic or hot melt, which aids in forming an environmental barrier for the splice. Diameters of the heavy wall tubing may be up to 7 in (178 mm) or even 12–24 in (300–600 mm) when it is used as a corrosion protection sleeve on weld joints of gas and oil pipelines [75, p. 248].

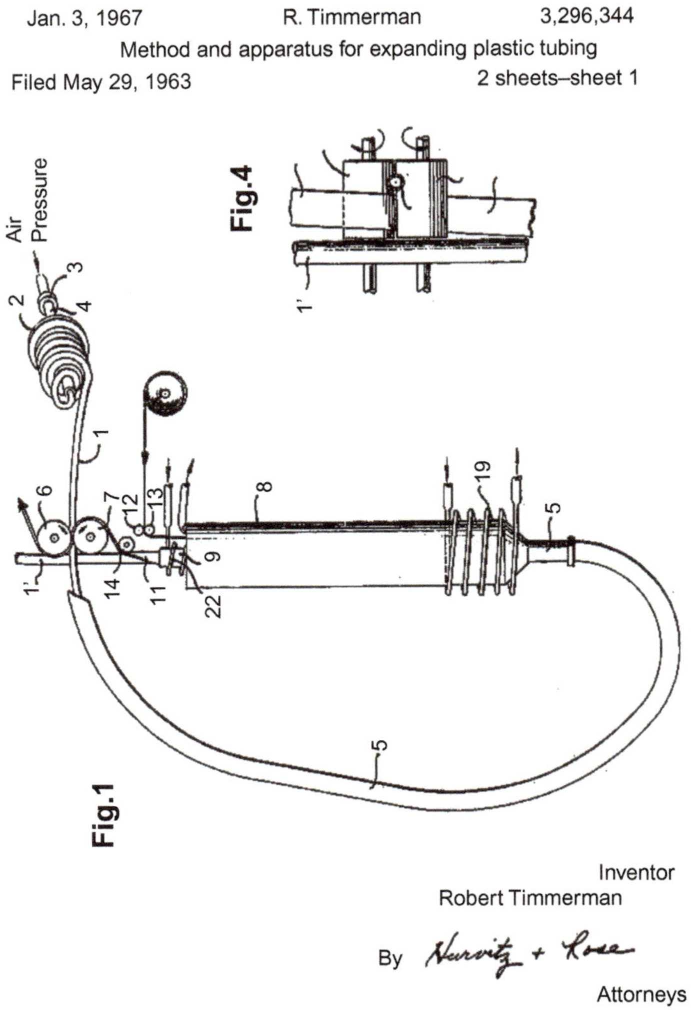

Figure 6.18 Alternative Patented Method of Producing Heat-Shrinkable Tubing [78].

6.4.2 Heat-Shrinkable Sheets and Films

Heat-shrinkable sheets (thickness 0.040–0.120 in, or 1–3 mm) and films (thickness 0.001–0.020 in, or 0. 025–0.5 mm) are fabricated from many of the same materials as shrinkable tubing [75, p. 248]. They are produced by extrusion as a tube, sheet, or blown film, and then irradiated. Orientation (stretching) after irradiation can be performed by several methods, namely by differentially heated and driven rolls (in the machine direction) or by a tenter frame (across the machine direction) as shown in Figure 6.20. If desired,

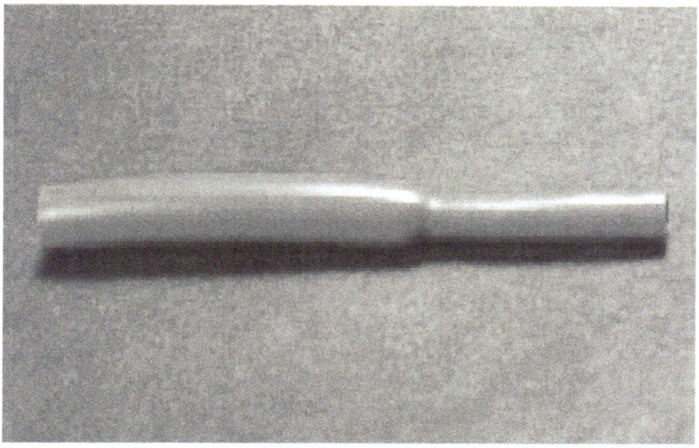

Figure 6.19 Plastic Memory Effect of Cross-linked Heat-Shrinkable Tubing. *(Courtesy of M.R. Cleland.)*

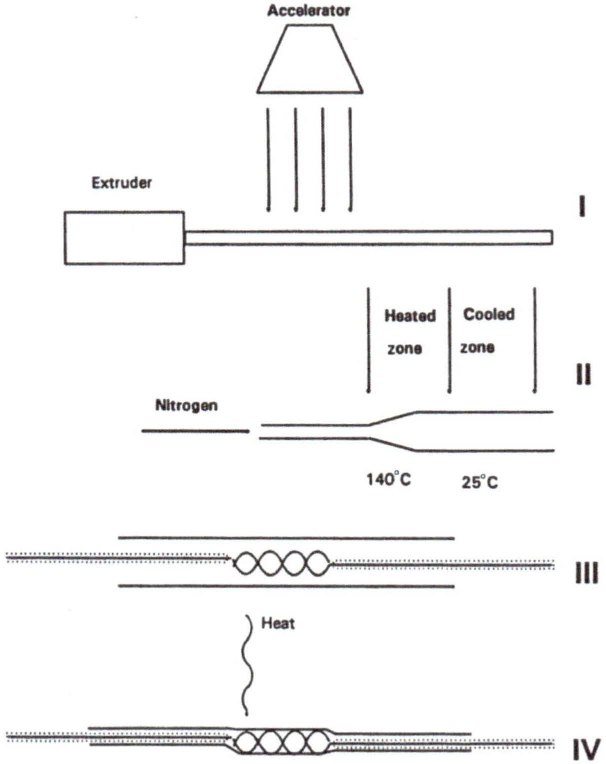

Figure 6.20 Schematic of Production and Application of Heat-Shrinkable Tubing onto a Wire Splice.

biaxial stretching may be accomplished by the combination of both procedures in succession or by a tenter frame specially designed for biaxial stretch (Figure 6.21).

Heat-shrink sheets are used for many of the same applications as heavy wall shrink tubing. The advantage of the sheet over a tube is that it can be conveniently slipped over the area to be protected, e.g., over continuously installed cable. There are many methods to close the heat-shrinkable sheets, such as zippers, rail and channel, and heat sealable bonds, some of them patented [80–83].

Heat-shrinkable films have found wide use in food wrap and packaging. The original development was done in the late 1950s by W.R. Grace and Co. and the process is still used for much more sophisticated multilayer laminated films with superior barrier properties [83–86], and exceptionally high tear strength [87], or for multilayer films with low shrink force for packaging of easily deformable articles [88]. The currently used doses for PE films are 200–300 kGy at electron energy in the range from 0.5 to 1.0 MeV [76]. The original Cryovac patent on heat-shrinkable film is shown in Figure 6.22 and

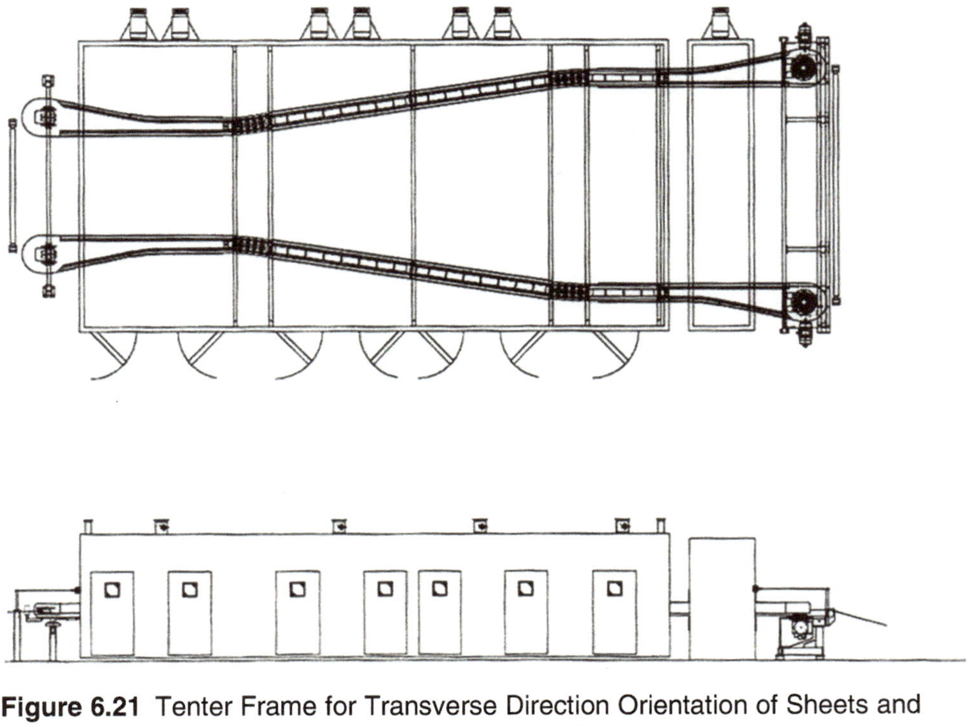

Figure 6.21 Tenter Frame for Transverse Direction Orientation of Sheets and Films. *(Courtesy of Marshall & Williams Plastics.)*

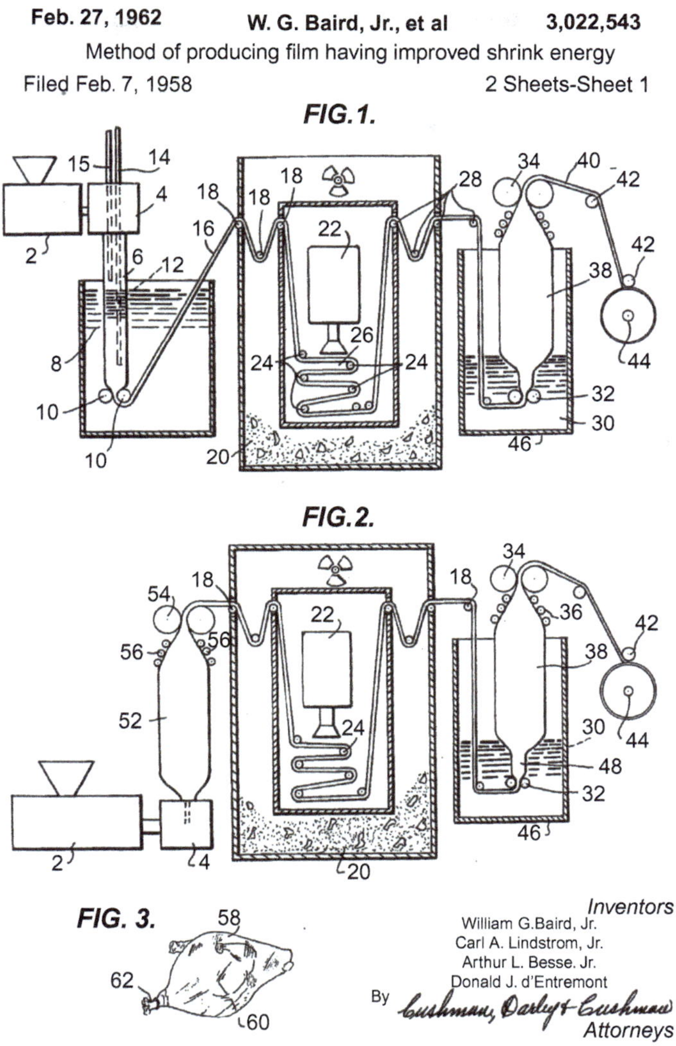

Figure 6.22 Original Cryovac Patent on Heat-Shrinkable Film (U.S. patent 3022543).

the currently operating line of 10 EB units for the large-volume manufacture of heat-shrinkable films is shown in Figure 6.23.

6.5 Cross-linked PE Pipes

Cross-linked PE pipes (PEX) have been used during the past several years as a replacement for copper and steel pipes mainly in plumbing and heating

Figure 6.23 Line of 10 EB Units for Heat-Shrinkable Films.

applications. The main advantage of PEX pipes over metal pipes is their light weight, resistance to corrosion, and scaling. Almost all PEX pipes are made of high-density PE (HDPE).

There are three primary methods for producing PEX pipes [89]:

1. The *Engel* or *peroxide method* employs a ram extruder (with a plunger action). In the extruder, peroxide is added to the base resin and through a combination of pressure and high temperature the cross-linking takes place as the tubing is produced.
2. The *Silane method* of PEX production involves grafting a reactive silane to the main chain of PE. The tube is produced by blending this grafted compound with a catalyst, which can be done using the *Sioplas* method or, by using a special extruder, the *Monosil* method. After extrusion, the tubing is exposed to either steam or hot water to induce the final cross-linking reaction in the tube.
3. *EB cross-linking* takes place when very high energy radiation is used to initiate cross-linking of the PE. This product is extruded like normal HDPE and then taken to an EB facility and routed under an EB accelerator. There it is exposed to a specific amount of radiation to achieve the required degree of cross-linking.

In European standards, these three methods are referred to as PEX-A, PEX-B, and PEX-C, respectively, and are not related to any type of rating system. All the resulting PEX tubing products perform similarly and are rated for performance by the ASTM, NSF, CSA, ISO, DIN, and other standards for which they are tested and certified. For example, the German DIN 16892 standard sets the following minimum degrees of cross-linking for the PEX materials [90]:

Material	Method of Cross-linking	Degree of Cross-linking, %
PEX-A	Peroxide	≥ 70
PEX-B	Silane	≥ 65
PEX-C	Irradiation	≥ 60

Cross-linking of HDPE pipes and tubing confers several important advantages:

- Improved heat resistance
- Improved environmental stress cracking resistance
- Improved pressure (stress) rupture resistance at elevated temperatures
- Improved chemical and swelling resistance
- Improved long-term strength at elevated temperatures
- Increased flexibility.

6.5.1 Irradiation of PE Pipes

Irradiation of PE pipes can be done by either EB or γ-rays. The latter method is more suitable for larger diameter pipes, where EB would not provide sufficient penetration. There are two methods of EB irradiation of the pipes: (1) static method and (2) dynamic method. The static method involves two-sided irradiation of the pipes and in the dynamic method rotating pipes are irradiated under the beam on a conveyor [45]. Typical conditions for EB cross-linking of HDPE pipes are given in Table 6.11 and comparison of one-sided and two-sided irradiation is given in Table 6.12. An example of performance data for EB irradiation of PE pipes is given in Table 6.13 [90].

Advantages of the EB irradiation method are as follows:

- High efficiency, high throughput
- Manufacturing cost savings

Table 6.11 EB Cross-linking of HDPE Pipes

Beam Properties and Performance	Electron Energy (MeV)			
	1.5	**2.0**	**2.5**	**3**
Beam current, mA	50	75–100	50	50
Beam power, kW	75	150–200		
Maximum tube diameter, in. (mm)	0.75	0.75	1.0	1.0
	(19)	(19)	(25)	(25)
Line speed, m/min (ft/min)	220 (67)	330 (100)	220 (67)	220 (67)

- Easy operating and maintenance
- Possibility of starting and stopping the process as required
- High reliability
- Uniform quality
- Automatic report generation during the operation
- Ecologically friendly, no extractable chemicals in the material.

PEX pipes are used mainly in plumbing and heating applications. The temperature and pressure in the usual applications, such as tubing, are as high as 95°C (203°F) and 10 bar (145 psi) [89,90]. For plumbing applications, the recommended maximum temperature and pressure are 180°F (82°C) and 100 psi (7 bar) [89]. Typical applications are hot water supply, space heating systems, service lines, hydronic radiant heating and cooling systems, snow melting applications, outdoor turf conditioning, residential fire-sprinkler systems, ice rink systems, and permafrost protection beneath refrigerated warehouses.

6.6 EB Process in Coatings, Adhesives, Paints, and Printing Inks

The main process in these applications is the direct conversion of liquids into solids. When using irradiation by either UV or EB, this conversion occurs almost instantly. There are specific areas where EB irradiation is more suitable than irradiation by UV light. In general, these include applications where thick layers of coatings or adhesives are applied. Other instances are coatings with high levels of inorganic pigments and/or fillers, which usually cannot be cured by UV radiation because of their opacity. As pointed

Table 6.12 Comparison of One-Side and Two-Side Irradiation of HDPE Pipes

Tube Size, mm^2 ($in.^2$)	Required Electron Energy, MeV[a]	
	Irradiation of One Side	Irradiation of Both Sides
14 × 1.5 (0.55 × 0.060)	2.5	1.44
16 × 2 (0.63 × 0.080)	3.05	1.27
17 × 2 (0.67 × 0.080)	3.16	1.31
18 × 2 (0.71 × 0.080)	3.26	1.36
20 × 2 (0.79 × 0.080)	3.46	1.44

[a]*Equal entrance–exit doses.*

Table 6.13 Performance Data for Irradiation of PE Pipes

Electron Energy, MeV	1.5	2	2.5	3
Beam current, mA	50	75–100	50	50
Beam power, kW	75	150–200	125	150
Application range up to tubing diameter, in. (mm)	¾ (19)	¾ (19)	1 (25)	1 (25)
Cross-linking speed, m/min	55	80	55	55

out earlier, the capital cost of standard EB curing equipment is considerably higher than that of a UV curing line. However, the recent trend is to build smaller EB processors operating at much lower voltage (see Section 3.2.1). Such machines are considerably less expensive and consequently represent a formidable competition to UV curing equipment in an increasing number of applications. At any rate, EB curing lines operate at much higher line speeds and compare favorably if they are used for continuous long runs. In this section, discussion will be limited to applications specific to the EB curing process or possibly a combination of EB and UV. Another important advantage of EB cure is that the inks used in this process do not contain any extractable substances, such as photoinitiators, which may be toxic.

6.6.1 Magnetic Media

Magnetic media, such as magnetic tapes for audio and video recording, program cards of various types, and computer diskettes, have an important place in the electronic industry. The magnetic particles, such as γ-ferric oxide, ferric oxide with Cr^{+2}, chromium oxide, and iron–cobalt alloys, are dispersed in a solution of a binder [91]. The binders include polyurethanes, vinyl chloride/vinyl acetate copolymers, vinyl chloride/vinyl acetate/vinyl alcohol copolymers, and polyacrylate nitrocellulose. Particularly important are blends of polyurethanes with other polymers [92]. The substrates used for tapes and other forms of magnetic media are PVC, PET, and other film-forming polymers, with PET being currently most widely used [93].

EB is very well suited for the production of most of the magnetic media, such as high-density floppy disks and magnetic tapes. In this process, the substrate is coated with a coating containing a large volume of magnetic particles. The uncured coating passes first through a magnetic field to orient the particles and then through the EB processor. The high loading of the magnetic particles excludes the use of UV cure. The EB offers several advantages: The high reaction rate virtually eliminates surface contamination closely associated with thermal cure. The high degree of cross-linking of the binders used in the coatings gives a tougher and more wear-resistant surface.

6.6.2 Coatings

Most of the EB units sold in the recent past were mainly for release coating applications. New chemistry has been commercialized, but further developments are needed to satisfy this market segment [92].

Another process, where EB equipment is widely used, is curing of overprint varnishes (OPV). OPV is applied over a printed surface to protect the ink layer and to improve the appearance of the product [92]. The exceptionally high gloss of such finishing coatings and varnishes is achieved specifically by EB cure [93,94].

Typical applications of EB curing in protective and decorative coatings include [94,95]:

- Wood finishes (doors, front panels of furniture, lamination and lacquer finishes, shelving, cabinetry, prefinished flooring, chairs, tables, guitars, broom handles, picture board frames, moldings).
- Paper coatings and finishes (paper laminates for wood decoration; high gloss paper for gift wrapping; overprint varnishes for book, magazine,

brochure covers; cereal, game, and frozen food boxes; liquid containers; labels; lacquer on metalized paper).

- Release coatings for textured casting papers, roll or sheet tables, tapes, hot melt adhesives and films, separating papers for chipboard panel presses [96].
- Film coatings (antistatic coatings, antiscratch coating on phone and other cards).
- Bindings for abrasive papers.

Coatings on *rigid substrates*, mainly on wood (doors, furniture, laminated panels), are often fairly thick (up to 200 g/cm^2) and consequently require electrons with relatively high energy. However, production rates are mostly dependent on other operations, such as feeding, discharge, and polishing; therefore low-to-moderate beam powers are sufficient. Curing of top lacquer on doors is shown in Figure 6.24 (Courtesy of Elektron Crosslinking AB).

Coatings on *flexible substrates* are considerably thinner, typically in the range 1–30 g/cm^2, but the production rates are high (100–300 m/min); therefore, lower energy electrons and moderate- to high-power beams are required [95].

An example of EB curing of coil or web coatings is shown in Figure 6.25 and that of curing line for the hybrid UV/EB process in Figure 6.26.

6.6.3 Printing and Graphic Arts

The objective of printing is to create a visibly identifiable image, consistently for a large number of impressions. In principle, this can be done with a printing plate and the various printing methods are named after the nature of the printing plate. Many techniques have been developed for this purpose. Flexography, lithography, and gravure are the main printing techniques, which account for the vast majority of printing applications. Each of these methods has a number of variations. Graphic arts are well-established application of radiation curing worldwide. Essentially, the processes used in graphic arts include the generation of images to be reproduced on to the printing plate, silkscreen, etc., and the use of radiation-curable inks and overprint varnishes. Many imaging processes rely on the exposure of materials to radiation to bring about the change in solubility in a solvent system (organic or aqueous), thus enabling exposed and unexposed areas to be differentiated. The differentiation between exposed and unexposed areas can lead to selective delamination, softening, tackiness (which may affect the adherence of toner powders) or a change in refractive index, which leads to holographic effects.

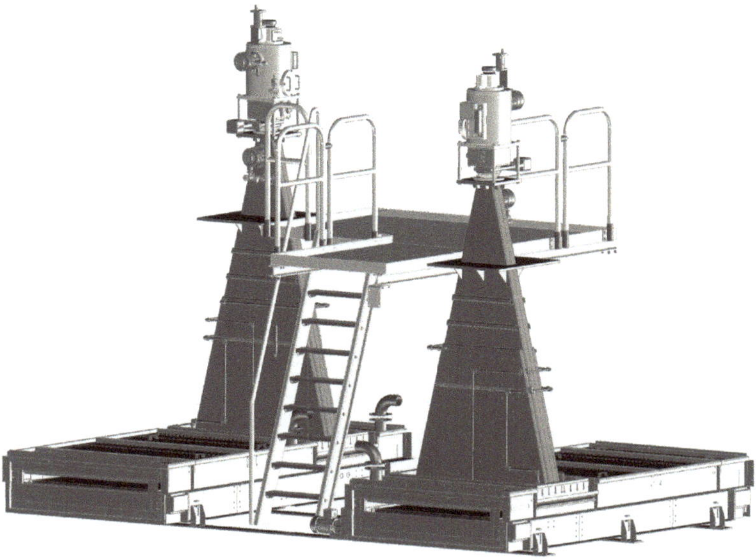

Figure 6.24 Curing the Top Lacquer on Doors. *(Courtesy of Elektron Crosslinking AB.)*

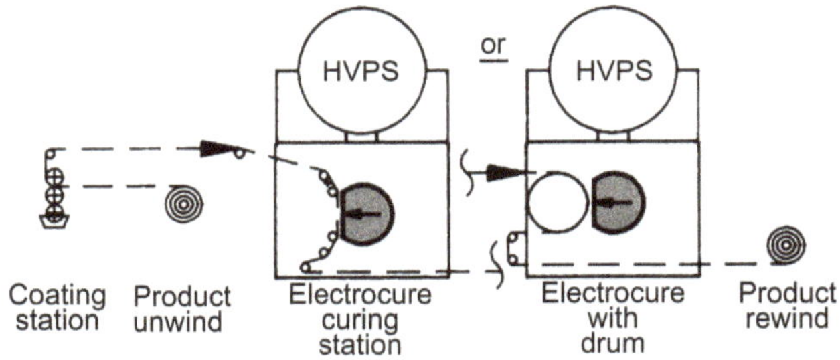

Figure 6.25 Schematic of Electron Beam Curing of Coil or Web Coatings. HPVS, High-Voltage Power Supply. The Arrow in the EB Unit Indicates the Direction of the EB. *(Courtesy of Energy Sciences Inc.)*

6.6.3.1 Flexography

Flexography is a mechanical printing process that uses liquid ink and a fairly soft relief image printing plate made of rubber, or more commonly photopolymer and pressure to create an image. Completed printing plates are mounted with adhesive onto a metal cylinder that rotates against the substrate during printing. A typical flexographic printing station is shown in Figure 6.27. A chambered doctor blade ink fountain applies ink to the engraved *transfer*, or *anilox, roll*. The

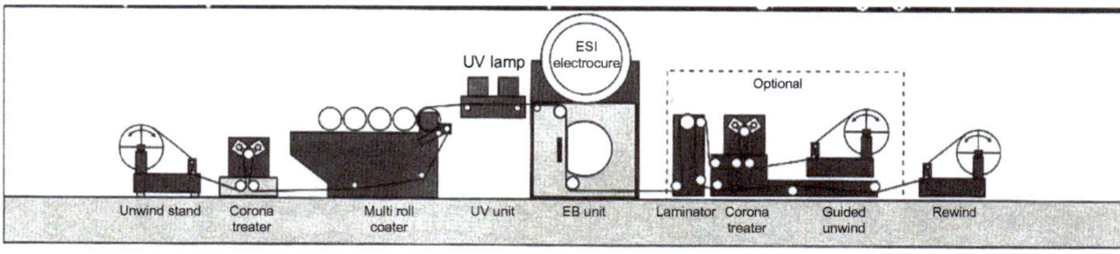

Figure 6.26 Schematic of Coating Line for a Hybrid UV/EB Curing Process for Coatings. (A) Unwind—Coating Section—UV Curing—EB Curing—Rewind OR: (B) Unwind—Coating Section—UV Curing—EB Curing—Optional: A Second Web Unwind and Lamination Station—Rewind. *(Courtesy of Faustel Corporation.)*

engravings on the anilox roll meter the correct amount of ink, depending on the engraving geometry and depth. The ink is transferred to the raised surfaces of the printing plate attached to the plate cylinder. The substrate is passed between the plate cylinder and the impression cylinder to achieve ink transfer.

The process is used mainly for package printing. All flexographically printed packaging substrates are web-fed with the exception of corrugated board. Unlike other major mechanical printing processes (namely lithography, gravure, and letterpress), flexography is in a phase of rapid development and change. Flexography is currently the major package printing process and it is very unlikely that this position will change over the next few years even in the light of digital printing technology.

The greatest strengths of using radiation cure in flexography are its versatility and cost-effectiveness. Since the inks do not dry until they have passed under the EB, the ink stays completely open. Unsightly print defects like dot bridging caused by dried ink on the printing plate are all but eliminated. In summary, main advantages of radiation-cured flexography over traditional technology are:

- Improved repeatability for printed spot colors
- High-definition process printing; better dot structure, lower dot gains
- Efficient job setup time.

6.6.3.2 Rotogravure Printing

Gravure printing separates printing from nonprinting areas by an engraved pattern that is chemically etched or mechanically cut into a surface of a hollow metal cylinder. Gravure cylinders are made from copper-plated steel. The printing pattern can be photographically imposed onto the cylinder surface and chemically etched or it can be imposed by a laser engraved with a stylus.

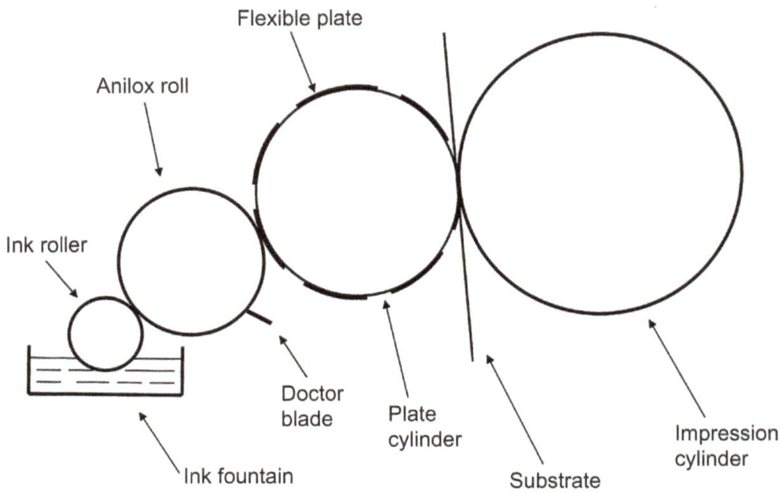

Figure 6.27 Flexographic Printing Deck.

Normal gravure printing is done from rolls and in web form. The entire surface of the gravure cylinder is flooded with low viscosity ink and then wiped clean with a straight edged doctor blade. The ink remains inside the recessed cell pattern. The substrate to be printed is nipped between the impression roll and the gravure cylinder and the ink is deposited on the substrate. Gravure printing is used for large-volume applications, such as labels, cartons, carton wraps, and flexible packaging materials. A rotogravure printing station is shown schematically in Figure 6.28.

Besides printing and decoration, flexographic, gravure, and lithographic plates as well as silk screens are used in resist chemistry, which is widely employed in the production of printed circuit boards. A resist is a material that will resist solvent attack. A negative resist is a material, which, upon exposure to light, cross-links and therefore becomes less soluble in a solvent system that would dissolve nonirradiated material. A positive resist is a material which becomes more soluble after irradiation due to depolymerization.

A typical lithographic station is shown in Figure 6.29.

With the advent of less expensive and more compact EB processors, a growing number of printers/converters in graphic arts are exploring the possibility of using this process as an alternative to conventional ink/coating drying [77].

Currently, the following EB cure applications in the printing and graphic arts are reported in the literature [94,96]:

- High gloss cosmetic and cigarette packaging
- Greetings cards

- Aseptic packaging
- Record album jackets
- Stamps
- Banknotes.

In some cases, a hybrid system (a combination of UV and EB cures) is used to ensure a sufficient degree of cross-linking through the coating and/or sufficient adhesion to the substrate. The additional benefit of such a hybrid system is the control of gloss and surface finish.

The most commonly found installation of EB equipment is at the end of a web offset press used for the production of folding cartons (Figure 6.30). Offset (lithographic) printing uses paste inks, which are designed to be "wet trapped" without any interstation drying. That allows placing a single EB unit at the end of the press [97]. The development of compact low-voltage EB equipment has allowed its use to cure inks on flexible packaging, polyester, and PP films and labels [97–101]. The low-voltage permanent vacuum modular equipment has a potential to be used for EB curing between the stations [102].

6.6.4 Adhesives

Adhesives are nonmetallic materials used to bond other materials, mainly on their surface through adhesion and cohesion. For adhesives working with cohesive phenomena, which is their majority, the adhesive fluid is transformed after bonding into a solid; this is typical for *laminating adhesives*

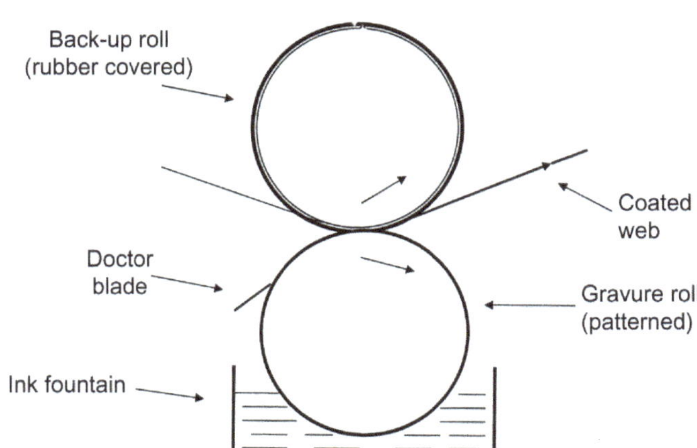

Figure 6.28 Rotogravure Printing Station.

(Section 6.6.4.2). In another scenario, the adhesive maintains its fluid state even after the bonding has occurred. Thus its debonding is moderate and the joint may be delaminated without destroying the laminate components in most cases. This is characteristic for *pressure-sensitive adhesives* (Section 6.6.4.1). There are other adhesive materials and techniques involving polymers, their solutions or dispersions, oligomers, and monomers. In the wide field of polymeric adhesives, EB curing has been used mainly in pressure-sensitive (PSAs) and laminating adhesives [103].

6.6.4.1 Pressure-Sensitive Adhesives

The main property that distinguishes a PSA from other types of adhesives is that it exhibits a permanent and controlled tack. This tackiness is what causes the adhesive to adhere instantly, when it is pressed against a substrate. After it has adhered, the PSA should exhibit tack, peel, and shear properties, which are reproducible within narrow limits. This requires that the adhesive layer be only slightly cross-linked [103]. PSAs are based on polymers with low T_g, typically in the range $-74°C$ to $+13°C$ [104].

Because of the precise control of the degree of cross-linking attainable by the EB process, it is well suited for the production of PSAs. Recent developments in EB design, particularly lower voltage and in materials

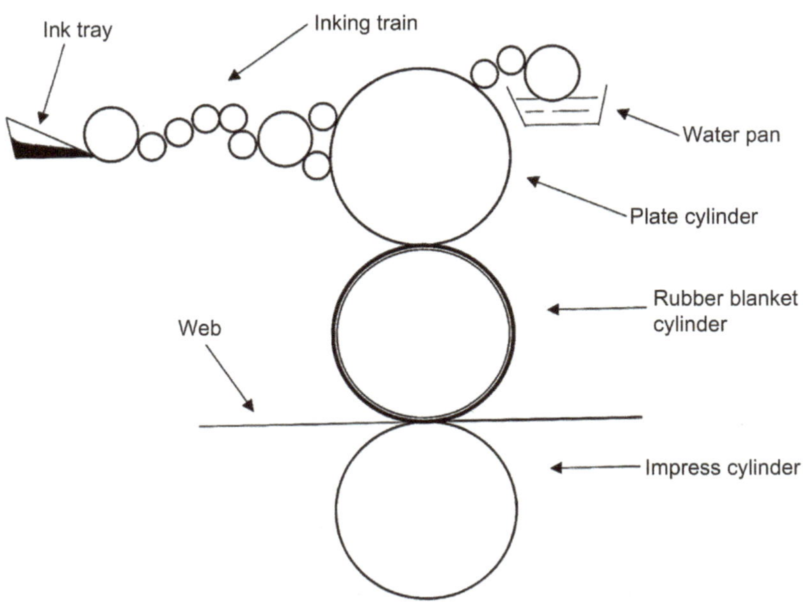

Figure 6.29 Lithographic Station. *(Courtesy of Foundation of Flexographic Technical Associations.)*

Figure 6.30 Low-Voltage EB Equipment on a Web Offset Press. *(Courtesy of PCT Engineering Systems LLC.)*

developments, have made EB curing of PSAs affordable, practicable, and possible [105–107].

Typical EB doses used for PSAs are 15–20 kGy [108]. The advantages of EB curing of PSAs are:

- High running speeds up to 900 m/min (3000 ft/min) are possible.
- Thermally sensitive films can be used.
- High coating weights are possible.
- No drying oven is necessary.
- Two-sided in-line coating is possible.

In general, EB curing of PSAs provides the following benefits to the manufacturer:

- New products can be developed by using a variety of substrates.
- A considerably higher productivity because of higher speeds and also because, typically, wider webs can be used.

- EB users are in compliance with environmental regulations because the systems use solventless adhesives, inks, and coatings.

The disadvantages are initial investment cost of the equipment and the necessity of being able to process long runs of the given product. In some cases, the prepolymer/diluting agent blend is highly viscous, so smootheners are necessary [109]. Another problem is the toxicity of some of the monomers used [110].

6.6.4.2 Laminating Adhesives

Laminating adhesives are used to bond layers of different materials together. There are many possible combinations, including different polymers, copolymers, oligomers, monomers, tackifiers, bonding paper, foils, fabrics, films, metal, and glass. The bond between the individual layers has to be sufficiently strong to hold the laminate together. This is accomplished by applying the adhesive in the molten state (hot melt) or in a solvent or aqueous dispersion and creating the bond. The bond can be created by cooling the melt or evaporating the solvent or water. In many cases, the bond is improved by cross-linking. Cross-linking can be accomplished by using a reactive adhesive, consisting of two components which react when mixed (e.g., polyurethanes or epoxies), or by radiation.

Cross-linking increases the cohesive strength of the adhesive and consequently the bond strength. Many of the laminations used are transparent to light and these can easily be cured by UV or visible light. However, if the laminated materials are opaque to UV and visible light, they may be cross-linked by EB since high-energy electrons penetrate paper, foils, and fabrics. Laminates of thin films or thin film overlays can be processed by low-energy EBs. For setting adhesive bonds between thicker substrates, higher energy or even X-ray radiation may be used. Materials with very different coefficients of expansion can be bonded by EB-curable adhesives without the interfacial stresses that are created when using thermal curing.

EB curing has been used for curing laminating adhesives in flexible packaging and this application has been growing rapidly since the introduction of low-voltage compact EB processors. The advantages of EB curing of laminating adhesives are as follows [111,112]:

- The adhesives used do not require solvents.
- The adhesive is one part chemistry (no mixing needed).
- Long shelf-life (more than 6 months).

- Adhesive remains unchanged until it is cured.
- No multiroll coating is needed.
- No complex tension controls are needed.
- The adhesive bond is established almost instantly.
- Real-time quality control.
- In-line processing; immediate shipment is possible.
- Easy cleanup.

Advantages over UV-curable laminating adhesives are as follows [112]:

- EB penetrates opaque films.
- EB produces typically higher conversion.
- Most EB systems cure without photoinitiators (simpler formulation, cost saving, fewer residues, and extractables).

Recently a solvent-free dual cure laminating adhesive based on the combination of polyurethane chemistry and EB cure was developed, which exhibits improved properties when compared to laminates prepared by either method alone [113].

6.7 Production of Fluoroadditives

Fluoroadditives, or PTFE micropowders, are finely divided low-molecular-weight PTFE powder. In general, they consist of small particles of the order of several μm. They are produced by several methods, namely degradation by heat, ionizing irradiation, and controlled polymerization. The raw material for the degradation methods is very often sintered or unsintered PTFE scrap, molding powders, production scrap, or postconsumer PTFE articles. The reduction of molecular weight and melt viscosity by degradation is illustrated in Table 6.14.

Table 6.14 Effect of Degradation on Properties of Polytetrafluoethylene Resins

Resin	Typical Molecular Weight	Typical Melt Viscosity, Pa s
Molding resin	10^6	10^{10}
Fluoroadditives (degraded resin)	10^4–10^5	10–10^4

The most economical way to produce fluoroadditives is a continuous process of irradiation of the resin and/or scrap under the EB source [114]. The processed material is spread on a conveyor belt in a specific layer thickness and is exposed to the required dose. Usually, the dose for sufficient degradation is 500 kGy or more. The advantage of multiple passes is that the material is allowed to cool down, which prevents it from sticking together and causing problems in the next step. The material heats up during irradiation, so the temperature is held at 121°C, which is sufficient to prevent sticking [111, p. 41]. Typical beam power of this process is at least 2.0 MeV. The degradation of the PTFE resin generates off-gases, such as hydrofluoric acid, which must be removed by ventilation of the processing area [111, p. 40].

The degradation process, i.e., reduction of molecular weight, decreases rapidly with the absorbed dose. A typical dose used for this process lies in the range 10–25 Mrad (100–250 kGy) [115]. When the irradiated resin has received a sufficient dose, it is ground to the desired particle size. The most widely found grinding method is employed in fluid energy mills, more commonly known as jet mills, in which jets of highly compressed air are used as the source of energy [116]. An example of a jet mill system is shown in Figure 6.31.

The ground micropowders have an apparent density of 400 g/l and a melting point of 321–327°C [111, p. 41]. Specific surface area ranges from 5 to 10 m^2/g and average particle size lies between 3 and 12 μm (Table 6.15).

PTFE micropowders are used as additives to lubricants (oils and greases), to plastics and elastomers to reduce friction and improve extrusion, to printing inks to reduce blocking and improve abrasion resistance, and to coatings to reduce wear and friction, and to increase water and oil repellency.

6.8 Radiation Curing of Polymeric Composites

6.8.1 Advanced Fiber-Reinforced Composites

Fiber-reinforced composites or advanced composites are materials combining high strength with a low weight. Polymeric composite materials consist of a thermoset or thermoplastic matrix and of reinforcing fibers. The main function of the matrix is to act as a binder for the fibers, to transfer forces from one fiber to the other, and to protect them from environmental effects and effects of handling. The reinforcing fibers can be continuous or discontinuous (short fibers) and may be oriented within the matrix at different angles. Short fibers are frequently dispersed in the matrix randomly, although some composite materials may contain oriented short fibers.

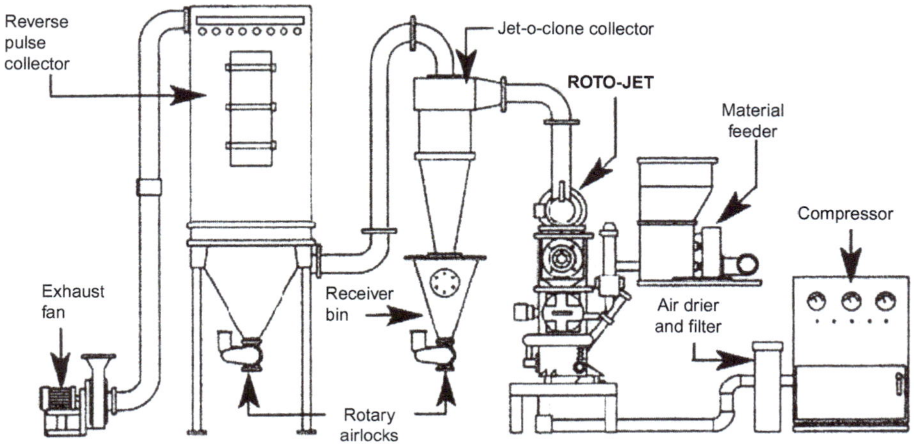

Figure 6.31 A ROTO-JET System Used for Grinding Irradiated Fluoropolymers. *(Courtesy of Fluid Energy Processing and Equipment Co.)*

Table 6.15 Comparison of EB and Chemical Cures in Wire and Cable Manufacture

EB Cure	Chemical Cures
No special extrusion equipment needed	Special extrusion equipment required
Noncritical extrusion conditions	Critical extrusion conditions
No heating media required	Hi pressure steam required
Instant start and stop	Slow start-up
Low scrap amount on start-up and stopping	High scrap amount start-up and stopping
High line speed	Low line speed
Wall thickness limitation	No wall thickness limitations
Compact, small space required	Large space requirement
Ozone and radiation are environmental and safety issues	Heat and steam are environmental factors

The performance characteristics of a composite material depend on the type of reinforcing fiber (its strength and stiffness), its length, fiber volume fraction in the matrix, and the strength of the fiber–matrix interface. The presence of voids and the nature of the matrix are additional but minor factors.

The majority of commercially available polymeric composites are reinforced by glass fibers, carbon fibers, aramid fibers (e.g., Kevlar) and to a lesser degree by boron fibers. In some cases, hybrid composites are made that contain combinations of fibers.

Matrix materials for commercial composites are mainly liquid thermosetting resins such as polyesters, vinyl esters, epoxy resins, and bismaleimide resins. Thermoplastic composites include polyamides, polyether ether ketone (PEEK), polyphenylene sulfide (PPS), polysulfone, polyetherimide (PEI), and polyamide–imide (PAI) and are used predominantly as melts.

Thermosetting composites are cured at either ambient or elevated temperatures to obtain a hard solid by polymerization and/or cross-linking. The use of radiation cross-linking decreases the cure time considerably. EB has been used in many instances successfully. For example, glass-fiber-reinforced composites cured by EB have been used for the production of cladding panels [117].

The production of materials having good mechanical properties matching those produced by conventional thermal methods has been achieved by using EB processing, and a growth in these applications is expected [118,119]. Graphite-fiber-reinforced composites with low stress and exhibiting little shrinkage upon cure have been produced with performance comparable to state-of-the-art toughened epoxy resins [120–122]. A layer-by-layer EB curing process of filament wound composite materials using low-energy EBs has been developed. This process has a wide range of applications in aerospace technology [123,124]. A variety of different curing methods leading to successful products is described in Refs [45, p. 309; 124, p. 312; 125, 126].

The article of Berejka [127] covers the state of the art in EB technology, material technology, and product-forming technology as applied to structural carbon-fiber-reinforced polymeric composites.

A great deal work was done to develop methods for using EB technology for the manufacture and repair of advanced composites. Many benefits have been identified for EB-curing fiber-reinforced composites, including lower residual stresses that result from curing at ambient or subambient temperatures, shorter curing time for individual components (minutes vs. hours), improved material handling (the resins have unlimited shelf life), and possible process automation in the placement of fiber reinforcement. The only process EB cannot be utilized in resin-transfer molding (RTM) since the EB cannot penetrate the massive molds used to mold the composite parts [128]. So far, in the main, EB-curable cationic initiated epoxy resins have been used. Although most of the development work has been geared to aerospace, the technology is applicable to other industries, including automotive and consumer goods.

The development work also yielded UV-curable adhesives for aluminum to aluminum, graphite to aluminum, and graphite to graphite.

So far, a number of parts have been produced by applying EB technology:

- Lower and upper wing assembly
- Liquid hydrogen tank
- Missile bodies (Figure 6.32)
- Composite armored vehicle
- Boat hulls
- Automotive body parts.

Another viable application is the use of EB curing in the repair of aircraft composite parts. It was determined that most such repairs are completed in 50–70% of the time required to complete a normal thermal repair [128].

The advantages of curing of advanced composites by ionizing radiation, which predominantly is done by EBs or X-rays, are summarized by Makuuchi and Cheng as follows [45, p. 305]:

1. *Lower energy consumption*: EB cure only uses 10% of the energy used for thermal curing.
2. *Shorter cure times that translate into a higher processing rate*: For example, a 200 kW and 10 MeV modern EB accelerator can provide a throughput of 2000–3000 kg/h (assuming 95% availability of the machine and 35–45% utilization of the beam energy).
3. *Ambient temperature curing*: Because of the much smaller temperature differences, the internal stresses in the composite microstructure are much smaller than for thermal curing.
4. *Possibility of making large products*: Very large products can be extremely expensive when cured by heat in autoclaves and can be cured with EBs at a much lower cost. The EBs can be converted into X-rays to overcome limitations in penetration.
5. *Simplified tooling*: Less complicated tooling can be used for ambient temperature curing and less costly materials can be used that are temperature sensitive.
6. *Improved material handling*: Ambient temperature curing reduces the need to take special precautions (e.g., refrigeration) to store uncured resins. Dissimilar materials and can be co-cured in a single curing

Figure 6.32 Composite Missile Body Irradiated by Electron Beam. *(Courtesy of A.J. Berejka.)*

cycle. There is also a possibility to reduce the number of processing steps.

7. *Better process control*: The degree of cure and location can be precisely controlled in radiation curing.
8. *Reduced VOC emission*: It has been estimated that VOC emissions can be reduced by up to 90% with radiation curing.
9. *Cost reduction is possible*: Although the initial capital investment for EB equipment is higher, overall curing cost can be lower due to higher throughput, lower energy consumption, less expensive curing reduction of VOC, and the number of processing steps.

6.8.2 Other Composite Materials

Polymeric composites can be *reinforced by natural fibers* such as hemp, flax, jute, banana, kenaf, pineapple leaf, and oil palm empty fruit bunch fibers [45, p. 321]. The advantage of using them is that they are renewable, more or less biodegradable, abundant, and inexpensive. They also have low density, relatively high strength, ease of separation, and do not cause excessive wear of process equipment as glass fibers do. Problems in using natural fibers, which are in general polar and hydrophilic, may occur if they are combined with nonpolar and hydrophobic matrix. Such reduced compatibility requires the addition of additives that increase the adhesion of the matrix and the surface of the fibers.

Experiments with EB-cured composites based on polystyrene reinforced by pineapple leaf fiber using acrylic prorads yielded a material with good mechanical properties [129]. Likewise a composite based on poly(ε-caprolactone), a biodegradable matrix reinforced by oil palm empty fruit bunch fiber, was cured by EB using polyvinyl pyrolidone as adhesion promoter and exhibited good mechanical properties and bond between matrix and reinforcing fiber [130].

Wood–plastic composites (WPCs) represent viable materials. This combination greatly reduces the shortcomings of wood: lack of dimensional stability based on its moisture content, flammability, poor resistance to chemicals, effect of weather conditions, biological attacks, and abrasion. WPCs may be produced by either chemical cure or radiation using ionizing sources. Radiation curing presents the advantage of not using chemical initiator and of being carried out at considerably lower temperature. The process has been used since the 1970s in North America. There are several methods available, such as using a variety of monomers (e.g., unsaturated polyesters, methyl methacrylate, styrene, vinyl acetate) [131,132], and or combination of wood and PP [133]. In the 1980s, the Dow Chemical Company developed and patented the process of producing a WPC by impregnating a wood substrate with liquid dicyclopentyl acrylate or methacrylate and curing the impregnated wood with ionizing radiation or by heating it in the presence of a catalytic initiator [134]. More recently, a process was developed in which a mixture of wood flour and LDPE with the addition of acrylate monomer was extruded and subsequently irradiated with EB at 80 kGy [135].

Wood-fiber-reinforced plastics (WFRPs) are composite materials where the wood fibers or other cellulosic fibers are dispersed in the plastic matrix as reinforcing fibers or fillers. In this case, special radiation-reactive adhesion promoters between the wood or cellulosic fibers have to be used to achieve a sufficiently strong bond between the polymeric matrix and the fibers. Examples of additives used for such a bond are maleic anhydride modified PP for wood-fiber-reinforced polyolefins and chemical coupling agents for fiber-reinforced thermosets, such as alkoxysilanes, silicon acrylates, and polymethylene polyphenylisocyanate [132]. Radiation processing of wood fibers to produce reinforced composites can be done by mutual irradiation or by preirradiation of the fibers only [132].

Laminar WPCs are essentially all plastic-coated or laminated paper products. They are frequently produced by radiation technology. Industrial-scale EB-cured wood panels ("boards") have been manufactured for a long time using EB accelerators with energy in the range of 150–300 keV. The main components of the EB-curable coatings in the laminates are multifunctional reactive oligomers that have unsaturated double bonds along the main chain or at the end of the chain. Examples are epoxy-acrylates or unsaturated oligoesters [132, p. 137]. Typical process conditions for boards of size 3.2 m × 1.5 m (10.5 ft × 5 ft) and 6–16 mm (0.25–0.63 in) thick are a conveyor speed of 5–15 m/min and dose rates 100–300 kGy/s and absorbed dose ranging 50–75 kGy [132, p. 142].

6.9 Hydrogels

Hydrogels are three-dimensional networks formed from hydrophilic homopolymers, copolymers, or macromers (preformed macromolecular chains) cross-linked to form insoluble polymer matrices. These polymers are synthesized by polymerization of hydrophilic monomers and are generally used above their glass transition temperature. They are typically soft and elastic due to their thermodynamic compatibility with water and have found use in a great variety of applications [136]. Synthetic monomers used for the preparation of hydrogels include, among others, ethylene glycol, vinyl alcohol, and polyacrylates such as 2-hydroxyethyl methacrylate (HEMA), *N*-isopropylacrylamide (NIPAM), and NVP [137]. The matrix polymers can be prepared and combined in the form of blends, copolymers, and interpenetrating networks (IPNs) [138].

Because of their nature, hydrogels are able to absorb large quantities of water. The extent of the reversible swelling and deswelling property of these materials is known to depend on the nature of both intermolecular and intramolecular cross-linking, as well as the degree of hydrogen bonding in the polymer network. Increasingly, these hydrogels are being utilized in a variety of applications including drug release, biosensors, tissue engineering, and pH sensors. To date, a variety of compounds have been utilized in the synthesis of hydrogels. The synthetic monomers mentioned above can be combined with natural polymers, such as agar and alginate [132, p. 3315].

The structures of the basic monomers used in the process, radiation techniques are very suitable tools for the production of hydrogels. Examples of the usage of a mature technology of this kind are for hydrogel wound dressings, being now produced on large scale [139], contact lenses (silicone hydrogels, polyacrylamide hydrogels), dressings for the healing of burn victims and other hard to heal wounds, disposable diapers, and similar highly absorbing products.

The most widely used hydrogels are based on PE oxide (PEO) dissolved at relatively low concentrations in water (typically 4–5%). Modest radiation exposure (less than 10 kGy) is needed to form the gel. Gels as thick as 2 mm (80 mils) are being produced.

Emerging applications include a hydrogel-based system in anticancer therapy for local drug delivery, systems for encapsulation of living cells, a new approach to the synthesis of polymeric material for intervertebral disk implantation, temperature-sensitive membranes, hydrogel phantoms of three-dimensional radiation dosimetry for radiotherapy, degradation-resistant nanogels, and microgels for biomedical purposes (e.g., synovial fluid substitute), hydrogel-based dietary products [139].

Another application reported [140] is preparing hydrogels from a mixture of polyvinyl alcohol (PVA) and carboxymethyl cellulose (CMC) using freezing and thawing and EB irradiation. A blend having a composition of 80/20 (CMC/PVA) has been used as a super absorbent in soil for agricultural purposes. Moreover, the water retention increased in the soil containing this hydrogel. Thus, this type of hydrogel can be used to increase water retention in desert regions.

Other hydrogels are prepared from poly(α-hydroxylic acids), e.g., lactic acid, polyacrylic acid, acrylamides and acrylated polyvinyl alcohol [141], and PE glycol.

6.10 Sterilization of Medical Devices

Sterilization of medical devices is a well-established process with the goal to reduce *bioburden*, which is normally defined as the number of bacteria living on a surface that has not been sterilized. In commercial sterilization, a quality assurance level (SAL) of 10^{-6} is required. This means that there is a probability of less than one in a million of any one article having any bioburden on it. In general, all types of ionizing radiation can be and are used for sterilization of medical devices and the choice of the radiation source depends on the dimensions and volume of the items to be sterilized. The issues involved in sterilization by ionizing radiation are (1) the materials used in the manufacture of the device; (2) when during the manufacturing operation will the device be exposed to the radiation sterilization; and (3) the amount or degree of exposure needed to attain the desired level of "sterility assurance" [142]. EB is used to sterilize a wide range of medical devices such as syringes, catheters, drains, tubing, urine bags, bandages, absorbents, gloves, surgical gowns and drapes, hand towels, culture tubes, and Petri dishes [143].

Ionizing radiation not only kills microorganisms but also affects the properties of materials used for the devices. For example, PVC that is often used to replace glass in some products can discolor upon exposure or the plasticizers used in it can leach into blood and other bodily fluids and may eventually become stiff or even brittle. In such cases, alternative materials, such as

PE blends containing optically clear metallocene catalyzed PEs or PPs and laminates derived from them, are being used [144]. Rigid transparent medical devices can be molded from radiation-tolerant polycyclic materials such as polystyrene, polycarbonate, and PE terephthalate or PBT [143].

EB equipment has been used for radiation sterilization since the 1950s. Typically, high-energy accelerators (10 MeV) have been used for several decades for the sterilization of packaged medical devices. Some mid-energy accelerators (3.0–5.0 MeV) are used for packaged medical devices with a low bulk density. Packaged products on pallets are irradiated by collimated X-rays using a patented process developed by MDS Nordion.

6.11 Other Applications for Ionizing Radiation

Other applications are proven and effective EB processes but are often limited by the size of a given market or by still developing commercial acceptance.

6.11.1 Battery Separators

EB cross-linking of PE films to exceedingly high cross-link density followed by surface grafting of acrylic acid by EB creates films that can control the ion flow between the anode and cathode of small lithium or other ion-based batteries [145–147]. These battery separators have significantly longer useful life than separator films produced by other methods [148].

6.11.2 Filter Membranes

Surface grafting is used to modify the hydrophilicity and hydrophobicity of filter membranes. For this purpose, microporous PVDF membranes are used that are grafted by monomers that are selected according to the desired end-product use [149].

6.11.3 Artificial Joints

Artificial hip and knee joints are made from radiation cross-linked UHMWPE. A normal hip joint unit consists of a stem, a ball, an insert, and a cup. A knee joint consists of a tibial component, a tibial insert, and a femoral component. The inserts are made of cross-linked UHMWPE [45]. The UHMWPE resins used for orthopedic applications have a molecular weight between 2 and 6 million with a degree of polymerization between 71,000 and 214,000. The stock materials of joint replacement components are

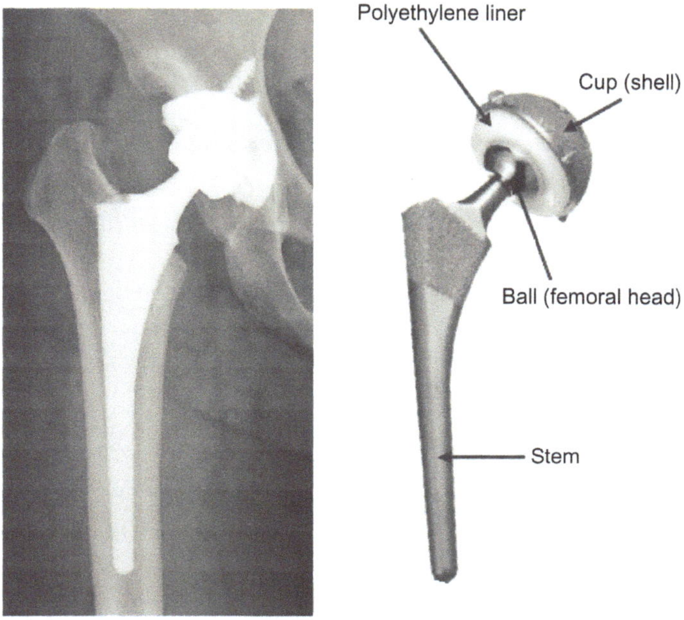

Figure 6.33 Example of an Artificial Hip Joint with an Insert from Irradiated UHMWPE.

manufactured from powders by ram extrusion or compression molding [147, p. 182]. The virgin resin exhibits relatively low wear resistance and long-term oxidation damage when used in joint replacement. The wear resistance is enhanced considerably by radiation cross-linking. Sufficient enhancement is attained by doses higher than 200 kGy. Postirradiation aging improves mechanical properties of the polymer [147, p. 186]. Oxidative degradation of UHMWPE during and after irradiation can be prevented by the addition of antioxidants. Vitamin E (tocopherol) was found to be an effective antioxidant for UHMWPE [150]. An example of an artificial hip joint using an irradiated UHMWPE insert is shown in Figure 6.33. The subject of the use of UHMWPE in orthopedic applications is covered in Chapters 6 and 12 of Ref. [45].

References

[1] Berejka AJ. Prospects and Challenges for the Industrial Use of Electron Beam Accelerators. International Atomic Energy Agency. Vienna, Austria. Publication SM/EB-01; p. 1. <www.iaea.org>

[2] Mehnert R, Bögl KW, Helle N, Schreiber GA. In: Elvers B, Hawkins S, Russey W, Schulz G, editors. Ullmann's encyclopedia of industrial chemistry, vol. A22. Weinheim: VCH Verlagsgesselschaft; 1993. p. 484.
[3] Burlant WJ. U.S. patent 3247012 (April 19, 1966) to Ford Motor Company.
[4] McGinnis VD. In: Mark HF, Kroschwitz JI, editors. Encyclopedia of polymer science and engineering, vol. 4. New York, NY: John Wiley & Sons; 1986. p. 445.
[5] Yongxiang F, Zueteh M. In: Singh A, Silverman J, editors. Radiation processing of polymers. Munich: Hanser Publishers; 1992. p. 72.
[6] Levy S, DuBois JH. Plastic product design handbook. New York, NY: Van Nostrand Reinhold; 1977.
[7] Rosen S. Fundamental principles of polymeric materials. New York, NY: John Wiley & Sons; 1982.
[8] Williams H. Polymer engineering. New York, NY: Elsevier; 1975.
[9] Roberts BE, Verne S. Plast Rubber Process Appl 1984;4:135.
[10] Hochstrasse U. Wire Ind 1985.
[11] Markovic V. IAEA Bull 1985.
[12] Brand ES, Berejka AJ. World 1978;179(2):49.
[13] Cleland MR. Radiat Phys Chem 1981;18:301.
[14] Barlow A, Hill LA, Meeks LA. Radiat Phys Chem 1979;14:783.
[15] Bly JH. Radiat Phys Chem 1977;9:599.
[16] Bruins PF. Plastics for electrical insulation. New York, NY: Wiley–Interscience; 1968.
[17] Harrop P. Dielectrics. London: Butterworth; 1972.
[18] Schmitz JV, editor. Testing of polymers, vol. 1. New York, NY: Wiley–Interscience; 1965.
[19] AEIC 5-79, Specifications, PE power cables (5-69 kV).
[20] ASTM D150, AC loss characteristics and dielectric constants. Philadelphia, PA: American Society of Testing Materials.
[21] Blythe AR. Electrical properties of polymers. London: Cambridge University Press; 1979.
[22] Wintle HJ. In: Dole M, editor. Radiation chemistry of macromolecules. New York, NY: Academic Press; 1972.
[23] Matsuoka S. IEEE Trans Nuclear Sci 1976;NS-23:1447.
[24] Sasaki T, Hosoi F, Haywara M, Araki K. Radiat Phys Chem 1979; 14:821.
[25] Barlow A, Briggs JW, Meeks LA. Radiat Phys Chem 1981;18:267.
[26] Chapiro A. Radiation chemistry of polymer systems. New York, NY: Wiley–Interscience; 1962.
[27] Salovey R, Gebauer R. J Polym Sci 1972;A-1, 10:1533.

[28] Waldron RH, McRae HF, Madison JD. Radiat Phys Chem 1988; 25:843.
[29] Peshkov IB. Radiat Phys Chem 1983;22:379.
[30] Vulcanization of Vistalon Polymers, ExxonMobil Chemical; 2000.
[31] Becker RC, Bly JH, Cleland MR, Farrell JP. Radiat Phys Chem 1979;14:353.
[32] Cleland MR. Technical information series, TIS 77-12. Melville, NY: Radiation Dynamics; 1977.
[33] Mc Laughlin WL, Boyd AW, Chadwick KH, McDonald JC, Miller A. Dosimetry for radiation processing. London: Taylor & Francis; 1989.
[34] Zagorski ZP. In: Singh A, Silverman J, editors. Radiation processing of polymers [chapter 13] Munich: Carl Hanser Verlag; 1992.
[35] Studer N, Schmidt C. Wire J Int 1984;17:94.
[36] Cleland MR. In: Singh A, Silverman J, editors. Radiation processing of polymers. Munich: Carl Hanser Verlag; 1992. p. 81.
[37] Luniewski RS, Bly JH. Irradiation and other curing techniques in the wire industry. In: Proceedings of the regional technical conference. Newton, Massachusetts; March 20–21, 1975, sponsored by Electrical and Electronic Division and Eastern Section of Society of Plastics Engineers.
[38] Barlow A, Biggs J, Maringer M. Radiat Phys Chem 1977;9:685.
[39] Clelland MR. Accelerator requirements for electron beam processing. Technical paper TIS 76-6. Melville, NY: Radiation Dynamics, Inc; 1976.
[40] Clelland MR. Lecture notes on electron beam processing. IAEA Regional Training Course, Shanghai; 1988.
[41] Timmerman. U.S. patent 3142629 July (1964) to Radiation Dynamics, Inc.
[42] Tada S, Uda I. Radiat Phys Chem 1983;22:575.
[43] Sawasaki T, Nojiri A. Radiat Phys Chem 1988;31:877.
[44] Hunt EJ, Alliger G. Radiat Phys Chem 1979;14:39.
[45] Makuuchi K, Cheng S. Radiation processing of polymer materials and its industrial applications. Hoboken, NJ: John Wiley & Sons; 2012. p. 155.
[46] Thorburn B, Hoshi Y. Meeting of rubber division of American Chemical Society. Detroit, MI; 1991 [paper #89].
[47] Thorburn B. Meeting of rubber division of American Chemical Society. Chicago, IL; 1994 [paper #20].
[48] Thorburn B, Hoshi Y. Meeting of rubber division of American Chemical Society. Anaheim, CA; 1997 [paper #89].
[49] Scherer WA. Radiat Phys Chem 1993;42:535.
[50] Charlesby A. Nucleonics 1954;12(6):18.
[51] Ivett RN. U.S. patent 2826540 (1958) to Hercules Powder Co.
[52] Precopio EM, Gilbert R. U.S. patent 2888424 (1959).

[53] Lewis JR. Jpn Pat Pub 1965;40-25:351 to Hercules Powder Co.
[54] Palmer DA. U.S. patent 3341481 (1967) to Hercules, Inc.
[55] Palmer DA. U.S. patent Re 26850 (1970) to Hercules, Inc.
[56] Scott HG. U.S. patent 3646155 (1972) to Midland Silicones, Ltd.
[57] Di-Cup® and Vul-Cup® Peroxides Technical Data; Bulletin PRC-102A. Wilmington, DE: Hercules, Inc; 1979.
[58] Griffin JD, Rubens LC, Boyer RF. Jpn Pat Publ 1960;35-13(138): or Br. Pat. 844-231 (1960) to Dow Chemical Co.
[59] Okada A, et al. U.S. patent 3542702 (1970) to Toray Industries, Inc.
[60] Atchinson G, Sundquist DJ. U.S. patent 3852177 (1974) to Dow Chemical Co.
[61] Nojiri A, Sawasaki Y, Koreda T. U.S. patent 4367185 (1983) to Furukawa Electric Co., Ltd.
[62] Lohmar E, Wenneis W. U.S. patent 4442233 (1984) to Firma Carl Freudenberg.
[63] Nojiri A, Sawasaki T. Radiat Phys Chem 1985;26(3):339.
[64] ASTM D2765, Method A, American Society for Testing Materials, vol. 08.02. Philadelphia, PA.
[65] Benning CJ. J Cell Plast 1967;3(2):62.
[66] Trageser DA. Radiat Phys Chem 1977;9:261.
[67] Harayama H, Chiba N. Kino Zairo 1982;2(10):30.
[68] Shina N, Tsuchiya M, Nakae H. Jpn Plast Age Dec. 1972;37.
[69] Sakamoto I, Mizusawa K. Radiat Phys Chem 1983;22(3-5):947.
[70] Shinohara Y, Takahashi T, Yamaguchi K. U.S. patent 3562367 (1971) to Toray Industries, Inc.
[71] Sagane N, et al. U.S. patent 3711584 (1973) to Sekisui Chemical Co., Ltd.
[72] Park CP. In: Klempner D, Frisch KC, editors. Polymeric foams. [chapter 9] Munich: Hanser; 1991. p. 211.
[73] Kiyono H, et al. U.S. patent 4218924 (1980) to Sekisui Chemical Co.
[74] Tamai I, Yamaguchi K. Jpn Plast Age 1972.
[75] Bradley R. Radiation technology handbook. New York, NY: Marcel Dekker; 1983.
[76] Cook PM, Muchmore RW. U.S. patent 3086242 (1963) to Raychem Corporation.
[77] Mehnert R, Bögl KW, Helle N, Schreiber GA. In: Elvers B, Hawkins S, Russey W, Schulz G, editors. Ullmann's encyclopedia of industrial chemistry, vol. A22. Weinheim: VCH Verlagsgesselschaft; 1993.
[78] Timmerman R. U.S. patent 3296344 (1967) to Radiation Dynamics, Inc.
[79] Silverman J. In: Singh A, Silverman J, editors. Radiation processing of polymers. Munich: Carl Hanser Verlag; 1992. p. 20 [chapter 2].

[80] Ellis RH. U.S. patent 3455336 (1969) to Raychem Corporation.
[81] Derbyshire RL. U.S. patent 4287011 (1981) to Radiation Dynamics, Inc.
[82] Muchmore RW. U.S. patent 3542077 (1970) to Raychem Corporation.
[83] Dyer DP, Tysinger AD, Elliott JE. U.S. patent 6326550 (2001) to General Dynamics Advanced Technology Systems, Inc.
[84] Brax HJ, Porinchak JF, Weinberg AS. U.S. patent 3741253 (1973) to W. R. Grace & Co.
[85] Mueller WB, et al. U.S. patent 4188443 (1980) to W.R. Grace & Co.
[86] Bornstein ND, et al. U.S. patent 4064296 (1977) to W.R. Grace & Co.
[87] Kupczyk A, Heinze V. U.S. patent 5250332 (1993) to RXS Schrumpftechnik Garnituren G.m.b.H, Germany.
[88] Bax S, Ciocca P, Mumpower EL. U.S. patent 6150011 (2000) to Cryovac, Inc.
[89] Plastic Pipe and Fitting Association, Glen Elyn, IL, <www.ppfahome.org> [accessed 25.05.12].
[90] Hoffman M. Presentation at the conference “Radiation processing for polymers of the 21st century”. Philadelphia, PA; April 8–9, 2003, sponsored and organized by IBA Advanced Materials Division, San Diego, CA.
[91] Koleske JV. Radiation curing of coatings. West Conshohocken, PA: ASTM International; 2002. p. 213.
[92] Santorusso TM. Radiat Curing 11(3);1983:4, Proceedings, Radcure’84, Atlanta, GA, September 10–13, 1984, p. 16 (1984).
[93] Zillioux RM. Proceedings of Radcure’86, Baltimore, MD; September 8–11, 1986, p. 8.
[94] Mehnert R, Pinkus A, Janorsky I, Stowe R, Berejka A. UV&EB curing technology and equipment, vol. 1. London/Chichester: SITA Technology Ltd/John Wiley & Sons; 1978.
[95] Maguire EF. RadTech Rep 1998;12(5):18.
[96] Seidel JR. In: Randell DR, editor. Radiation curing of polymers. London: The Royal Society of Chemistry; 1987. p. 12.
[97] Biro DA. RadTech Rep 2002;16(2):22.
[98] Gamble AA. In: Randell DR, editor. Radiation curing of polymers. London: The Royal Society of Chemistry; 1987. p. 76.
[99] Lapin SC. RadTech Rep 2008;22(5):27.
[100] Meij R. RadTech Europe 2005. In: Conference proceedings. Barcelona; 2005.
[101] Sanders R. Flexible packaging printer discovers EB technology. RadTech Rep 2003;17(5):30.
[102] Chrusciel J. Proceedings RadTech Europe 2001. Switzerland: Basel; 2001.

[103] Fisher R. In: Conference proceedings, TAPPI hot melt symposium 1999. Durango, CO; 1999, p. 115.
[104] Miller HC. J Adhes Sealant Council, Fall 1998. Conference proceedings, Chicago, IL; 1998, p. 111.
[105] Ramharack R, et al. In: Proc RadTech North America; 1996, p. 493.
[106] Nitzl K. European Adhesives Sealants 1996;13(4):7.
[107] Ramharack R, et al. Adhes Age 1996;39(13):40.
[108] Nitzl K. Adhesion 1987;6:23 [in German].
[109] Kupfer GA. Coating 1985;12:258.
[110] Der Polygraph (5), p. 366 (1986) [in German].
[111] Rangwalla I, Maguire EF. RadTech Rep 2000;14(3):27.
[112] Lapin SC. RadTech Rep 2001;15(4):32.
[113] Henke G. Proceedings RadTech Europe 2001. Switzerland: Basel; 2001.
[114] Ebnesajjad S, Morgan RA. Fluoropolymer additives. Oxford, UK: Elsevier; 2012. p. 39.
[115] Dillon JA, U.S. patent 3766031 (1973) to Garlock, Inc.
[116] ROTOJET: Size Reduction Systems, Fluid Energy Processing and Equipment Co., Hatfield, PA; 2008.
[117] Chaix C. RadTech Rep 1997;11(1):12.
[118] Walton TC, Crivello JV. Conference proceedings,'95 international conference on composite materials and energy. Montreal, Canada; 1995. p. 395.
[119] Singh A, et al. Conference proceedings,'95 international conference on composite materials and energy. Montreal, Canada; 1995, p. 389.
[120] Walton TC, Crivello JV. Materials challenge-diversification and the future. Volume 40: Book 2. Symposium proceedings. Anaheim, CA; 1995, p. 1266.
[121] Guasti F, Matticari G, Rossi E. SAMPE J 1998;34(2):29.
[122] Guasti F, Rossi E. Compos Part A: Appl Sci Manufact 1997;28A (11):965.
[123] Raghavan J, Baillie MR. Conference proceedings, polymer composites'99. Quebec, Canada; 1999, p. 351.
[124] Crivello JV, Walton TC, Malik R. Chemistry of materials 1997;9 (5):1273.
[125] Hill S. Mater World 1999;398.
[126] Beziers D, et al. U.S. patent 5585417 (1996) to Aerospatiale Societe Nationale Industrielle.
[127] Berejka AJ. Electron beam-cured composites: opportunities and challenges. RadTech Rep 2002;16(2):33.
[128] Lopata VJ, Sidwell DR. Electron beam processing for composite manufacture and repair. RadTech Rep 2003;17(5):32.

[129] Siregar J, Sapuan S, Rahman M, Zaman H. Paper presented at the 9th Malaysian national symposium on polymeric materials. Uniten; 2009.
[130] Ibrahim N, Ahmad S, Yunus W, Dahlan K. eXpress Polym Lett 2009;3:226.
[131] Czvikovski T. Radiat Phys Chem 1985;25:439.
[132] Czvikovski T. [chapter 7] In: Singh A, Silverman J, editors. Radiation processing of polymers. Munich: Hanser Publishers; 1992.
[133] Czvikovski T. Radiat Phys Chem 1996;47:425.
[134] Broxterman WE, et al. U.S. patent 4307155 (1981) to The Dow Chemical Co.
[135] Harper D, et al. Paper presented at the 9th international conference on wood and biofiber plastic composite, Madison, WI; 2007.
[136] Peppas NA. Crystalline radiation cross-linked hydrogels of poly(vinyl alcohol) as potential biomaterials. PhD thesis. Cambridge, MA: Massachusetts Institute of Technology; 1973.
[137] Bhattacharyya D, Xu H, Desmukh R, Timmons RB. Chem Mater 2007;19(9):2222.
[138] Slaughter BV, et al. Adv Mater 2009;21:3307.
[139] Rosiak JM, Janik I, Kadlubowski S, Kozicki M, Kujawa P, Stasica P, Ulanski P. Nucl Instrum Meth B 2003;208:325.
[140] El Salmawi KM. J Macromol Sci Part A 2007;44(June):619.
[141] Davidson RS. Radiation curing, Report 136, vol. 12, No. 4. Rapra Technology Ltd.; 2001. p. 27.
[142] Industrial Radiation Processing with Electron Beams and X-rays. International Atomic Energy Agency. Vienna, Austria; 2008. p. 65.
[143] Berejka AJ, Kaluska I. Materials used in medical devices. Trends in radiation sterilization of health care products. Vienna, Austria: International Atomic Agency; 2008. p. 159.
[144] Portnoy RC. Paper "Clear, radiation sterilizable, autoclavable blends based on metallocene catalyzed propylene homopolymer" presented at SPE-ANTEC. New York, NY, Proceedings, vol. 3. 1999. p. 3011.
[145] Machi S, et al. U.S. patent 5137137 (1979) to Japanese Atomic Energy Research Institute and Maruzen Oil Co., Ltd.
[146] D'Agostino, et al. U.S. patent 4230549 (1980) to RAI Research Corporation.
[147] Machi, S, et al. Paper presented at the second international meeting on radiation processing. Miami, FL; 1978.
[148] Industrial Radiation Processing with Electron Beams and X-rays. International Atomic Energy Agency. Vienna Austria; 2011, p. 77.
[149] Degen PJ. et al. U.S. patent 5282971 (1994) to Pall Corporation.
[150] Sakuramoto I, et al. The effects of oxidative degradation on mechanical properties of UHMWPE for artificial knee joint. J Soc Mater Sci Jpn 2001;67:1702 [in Japanese].

Recommended Further Reading

Makuuchi K, Cheng S. Radiation processing of polymer materials and its industrial applications. Hoboken, NJ: John Wiley & Sons; 2012.

Industrial Radiation Processing with Electron Beams and X-rays. Vienna, Austria: International Atomic Energy Agency; 2011.

Cleland Mr, Galloway, RA. Electron beam crosslinking of wire and cable insulation, technical information series TIS 01812 IBA industrial-white paper. Edgewood, NY: IBA Industrial.

Drobny JG. Radiation technology for polymers. Boca Raton, FL: CRC Press; 2010.

Slaughter BV, Khurshid SS, Fisher OZ, Khademhosseini A, Peppas NA. Hydrogels in regenerative medicine. Adv Mater 2009;21:3307.

Trends in Radiation Sterilization of Health Care Products. Vienna, Austria: International Atomic Agency; 2008.

L'Anunziata MF. Radioactivity: introduction and history. Amsterdam: Elsevier; 2007.

Benedek I. Pressure sensitive adhesives and applications, 2nd ed, New York, NY: Marcel Dekker; 2004.

Mehnert R, Pinkus A, Janorsky I, Stowe R, Berejka A. UV&EB curing technology and equipment, vol. 1. London/Chichester: SITA Technology Ltd/ John Wiley & Sons; 1998.

Kudoh H, Sasuga T, Seguchi T. High energy irradiation effects on mechanical properties of polymeric materials. Radiat Phys Chem 1996;48(5):545.

ACS symposium series 620, Clough RL, Shalaby WS, editors. Irradiation of polymers, fundamentals and technological applications. Washington, DC: American Chemical Society; 1996.

Singh A, Silverman J, editors. Radiation processing of polymers. Munich: Carl Hanser Verlag; 1992.

Yongxiang F, Zueteh M. In: Singh A, Silverman J, editors. Radiation processing of polymers. Munich: Carl Hanser Verlag; 1992. p. 77.

ACS symposium series 475, Clough RL, Shalaby WS, editors. Radiation effects on polymers. Washington, DC: American Chemical Society; 1991.

Taniguchi N, Ikeda M, Miyamoto I, Miyazaki T. Energy-beam processing of materials. Oxford: Clarendon Press; 1989.

Bradley R. Radiation technology handbook. New York, NY: Marcel Dekker; 1984.

Charlesby A. Radiation effects in materials. Oxford, UK: Pergamon Press; 1960.

7 Radiation-Resistant Polymers and Their Applications

OUTLINE

7.1 Degradation and Stability

Long-term degradation of materials exposed to ionizing radiation during their use is of vital concern. Relevant applications include both postexposure effects in materials subjected to radiation processing (such as cross-linking, grafting, chain scission, or sterilization) and in instances in which materials receive low-level radiation exposure throughout their useful life, as in a nuclear plant or space vehicles. In some cases, such as in space vehicles, they are exposed to radiation mainly in the absence of oxygen. However, in most applications, they are exposed to air, either during irradiation or subsequent radiation exposure. The effects of irradiation under oxidizing conditions can be very different from those occurring under an inert atmosphere.

Polymeric materials exhibit a wide range of radiation stabilities. Radiation resistance is strongly influenced by several factors, including the basic molecular structure, the presence of certain types of additives, and the particular environmental exposure conditions.

7.1.1 Radiation-Induced Degradation in the Absence of Oxygen

Scission and cross-linking are the most important chemical reactions leading to changes in mechanical properties of polymers undergoing radiation. Cross-links or chain breaks averaging only one per polymer molecule (e.g., involving only 1 out of 30,000 CH_2 units in PE of molecular weight 4.2×10^5) can have

Drobny: Ionizing Radiation and Polymers.
DOI: http://dx.doi.org/10.1016/B978-1-4557-7881-2.00007-9

profound affect on polymer properties. Polymers typically undergo simultaneous scission and cross-linking, but in most cases one or the other predominates [1]. Polymers subjected to degradation in the absence of oxygen can be grouped into two categories, in terms of the predominance of scissions versus cross-linking [2]. A very useful rule of thumb for predicting whether cross-linking or scission prevails holds for a large number of polyolefins: Polymers with a high concentration of quarternary carbon atoms along the chain, such as polyisobutylene, PMMA, and poly(α-methyl styrene), undergo predominantly scission, while polymers lacking this structural feature, such as PE, polystyrene, and NR, undergo primarily cross-linking. In most cases, polymers with predominant scissions exhibit considerable changes in their physical properties at lower absorbed doses in comparison with polymers that predominantly cross-link. Therefore, when selecting polymers as materials for the use in radiation applications, the polymers with dominating chain scission are generally avoided.

The presence of aromatic rings in the polymer chain has a strong stabilizing influence on the yield of radiation-induced cross-linking or scission due to the resonant energy mechanisms, and polymers bearing aromatic functionality are particularly useful in applications where radiation resistance is required. For example for PE, the cross-link yield, $G(X)$, is 1.0–2.5, for polystyrene, it is 0.035–0.050 [3,4]. For polyisobutylene, the $G(S)$ value is 0.25 [5] and for methyl styrene, the $G(S)$ value is 0.25 [6]. Likewise, the $G(X)$ in polyphenylmethylsiloxane is significantly lower [7] than that of PDMS and aromatic polyamides are more resistant to radiation than aliphatic polyamides [8]. Radiation resistance is not necessarily correlated with the resistance to other degradative effects. Thus PTFE ("Teflon") exhibits an outstanding thermal chemical resistance, yet it degrades rapidly when exposed to ionizing radiation [9].

7.1.1.1 Unsaturation and Discoloration

Another important structural change resulting from irradiation is the formation of unsaturated sites. This occurs due to the loss of adjacent substitutes on the chain or side group, such as adjacent hydrogen atoms in PE, or to radical–radical termination reactions that occur by disproportionation. Double bonds may subsequently participate in cross-linking by radical addition [1, p. 676]. *Cis–trans* isomerization of double bonds in unsaturated polymers such as polybutadiene or polyisoprene is also induced by ionizing radiation [10,11].

Some polymers undergo color changes when irradiated, because of the formation of conjugated double bonds or trapped radicals. PVC darkens at doses 50–150 kGy and many polyolefins tend to yellow at doses typically 100 kGy.

Polystyrene and polysiloxanes resist such color changes [1, p. 676]. Small amounts of impurities or additives (e.g., phenolic antioxidant stabilizers) promote markedly color formation in irradiated polymeric materials [10].

7.1.1.2 Formation of Gaseous Products

Another phenomenon related to radiation degradation of polymers is the formation of gaseous products. Irradiation of polyolefins produces mainly H_2 together with lower alkanes and alkenes (methane, ethane, ethylene, propane, and others) [11]. Other polymers produce a different mixture of gases and this depends on the atomic composition and molecular structure of the given polymer, for example. PMMA produces 18% hydrogen, 15% methane, 36% carbon monoxide, 25% of carbon dioxide, and 5.3% of propane [12]. Halogenated polymers generate highly corrosive gaseous products. For example, PVC has a high yield of HCl [13].

The volumes of gaseous products are substantial, e.g., PE irradiated by 1 MGy produces about 10 cm^3 gaseous products at STP from 1 g of polymer [1, p. 678]. These large volumes may become an issue when the irradiation is done in sealed containers. Moreover, there is a chance of formation bubbles in the irradiated material, causing either porosities or dimensional deformation.

7.1.1.3 Radiation-Induced Conductivity

Radiation-induced electrical conductivity is found in all organic polymers due to the generation of ions and the induced current is a function of the dose rate. The conductivity usually decays exponentially over periods of days or months, depending on the type of material [14]. Polymers can be made more resistant to induced electrical conductivity by doping with low concentrations of electron donors or acceptor molecules, which serve as deep, extrinsic trap sites [1, p. 683].

7.1.1.4 Radiation Resistance of Specific Polymers

Polystyrene, because of its high aromatic content, has the highest radiation resistance of any widely used inexpensive thermoplastic materials. The yield of gaseous products, which is mostly hydrogen, is 100 times lower than for aliphatic olefins. Polystyrene undergoes slow cross-linking when irradiated under an inert atmosphere with mechanical properties nearly unchanged at 10^4 kGy [15,16]; excessive damage to the material occurs only above 5×10^6 kGy. Copolymers of styrene, such as acrylonitrile butadiene styrene rubber (NBR) also exhibit a very good radiation resistance [1, p. 684]. Polyvinylchloride tolerates exposures up to 10^2 kGy without any significant change of mechanical properties.

Other, less common polymers, such as aromatic PI (e.g., Kapton) and poly (phenylene sulfide), are highly radiation resistant with their mechanical properties unchanged at doses up to 10^4 kGy [17,18]. Mechanical properties of PEEK and poly(*m*-phenylisophthalamide) are reduced by less than 50% above 10^4 kGy [18].

Many highly cross-linked thermoset resins, including phenolics, epoxies, and polyurethanes, are unchanged when exposed to very high doses; their radiation resistance is markedly improved by the addition of mineral or glass fillers [1, p. 686].

Several elastomers including polyurethanes, SBR, and some formulated ethylene–propylene elastomeric materials are highly radiation resistant. As for fabrics, aromatic polyamides (e.g., Kevlar) and PET exhibit much better radiation resistance than synthetic and natural fibers [1, p. 687]. PTFE, butyl rubber, and poly(oxymethylene) show low resistance with appreciable loss of useful properties at 100 kGy and below [1, p. 687].

7.1.2 Radiation Degradation in Air

Degradation by ionizing radiation is greatly affected if it is done in air, under these conditions, where extensive oxidation can occur inside the material. In such cases, oxygen becomes heavily involved in the degradation chemistry [1, p. 687]. The outcome in the presence of air is quite different from that under the oxygen-free atmosphere because the oxygen strongly favors chain scission. The presence of oxygen enhances the degradation of many polymers and its extent, and that holds whether it is during or after irradiation [1, p. 687].

Another factor, *dose rate*, can have a major effect on degradation of polymers irradiated in the presence of oxygen. Lower rates often increase oxidation and chain scission yields and result in more extensive material degradation per equivalent absorbed dose [1, p. 692]. Dose-rate effects are one of several mechanistically related, time-dependent phenomena that take place during radiation oxidation; others include temperature and postirradiation effects. Dose-rate effects occur due to oxygen diffusion, peroxide breakdown, and radical migration and have dramatic practical consequences [1, p. 692]. Examples are pipes made from LDPE and power cables failing in much shorter times when irradiated at low dose rates [1, p. 692].

Degradation of irradiated materials often continues, in storage or in use, long after the material has been removed from the radiation environment [19,20]. The postirradiation reaction is initiated by reactive intermediates formed during the irradiation [1, p. 693]. Significant *postirradiation oxidation effects* may take place over weeks to years, and may result in degradation that is much more extensive than that which was the immediate effect of the irradiation. The occurrence and magnitude of postirradiation effects are strongly material dependent. Materials free of crystalline regions that are effectively stabilized against oxidation are considerably less prone to postirradiation effects [1, p. 694].

Temperature influences radiation degradation in the presence of oxygen. Strong temperature effects can arise caused by change in key processes that affect the oxidation, such as peroxide decomposition, radical migration, oxygen diffusion, and diffusive loss of stabilizing additives [1, p. 694].

7.1.2.1 Stabilizers for Radiation Degradation in Air

Stabilizers are additives that improve degradation of polymers irradiated in air. Stabilizers are frequently antioxidants, which are mostly *radical scavengers* that interrupt the radical-mediated oxidation chain reaction. Antioxidants inhibit oxidative degradation occurring during and after irradiation. The stabilizing effect can be larger than that achieved in inert-atmosphere irradiation [1, p. 695].

Many antioxidants are effective in increasing the stability of polymeric materials irradiated in air and are often commercially available compounds [8]. Hindered phenols and amines are among the most effective and most widely used [1, p. 695]. Examples of typical antioxidants and their effects are given in Table 7.1. The most effective stabilizer, *N*-cycloxyl-*N*′-phenyl-*p*-phenylene diamine is reported to reduce the scission rate by a factor of 10 [1, p. 695].

Table 7.1 Dose Required for Reducing Tensile Strength to Half the Initial Value, for PE-Containing Stabilizers at 0.25%

Stabilizer	**Dose (kGy)**
None	6
2-Mercaptoimidazole (MBI)	6
Trilauryl phosphate	6
Ionox 330 [2,4,6-tri(3′5′di-*tert*-butyl-4-hydroxybenzyl) mesitylene][1]	8
2-Mercaptobenzothiazole (MBT)	13
DNPD [*N*-*N*′-di(β-naphthyl-*p*-phenylenediamine)]	15
Santonox R [4-4′-thiobis(6-*tert*-butyl-3-methylphenol)]	23
Santowhite powder [4-4′-butylidenebis(3-methyl-6-*tert*-butylphenol)][a]	24

[a]*A hindered phenol derivative.*

Stabilization depends on the content of the stabilizer up to a certain threshold beyond which there is little or no effect. More on this subject is given in Section 4.3.

7.1.3 Radiation Resistance of Specific Polymers

Approximate radiation resistance of particular polymer types can be defined as radiation degradation under inert (or very high dose-rate) conditions. Many polymers exhibit much less radiation resistance under oxidizing conditions, as discussed in Section 7.1.2. Thus there is no simple way to draw a conclusion from available data. The effect of changing from irradiation at a low dose under inert conditions to irradiation at a low dose in air depends on the material. For such polymers, such as silicone rubber, oxidation is not effective and the change in radiation resistance is small. For other polymers, such as PTFE or polystyrene, the factor is considerable; the radiation resistance is reduced up to a factor of 40 [1, p. 699]. That hard, glassy, and highly cross-linked materials show little difference between irradiation under inert conditions and irradiation in air at moderate to low dose is, in part, due to a low oxidation rate resulting from a very low rate of oxygen permeation. The comparisons are further complicated by dose rate, formulation sample thickness, temperature, and other parameters [1, p. 701]. In spite of all of the complicating factors, Table 7.2 is able to illustrate the irradiation resistance of selected polymers based on the dose required to reduce the elongation at break to 50% of the original. The data show the differences in these polymers irradiated under low dose rate in air and under high dose rate in inert atmosphere. The data are taken from Table 3 from Ref. [1, p. 700].

7.2 Applications of Radiation-Resistant Polymers

Several high-technology industries require specialty polymers that exhibit a specific response to exposure to radiation. For example, the electronics industry requires materials that undergo radiation-induced scission or cross-linking for resist application, while aerospace, medical applications, and the nuclear industry require highly radiation-stable materials.

7.2.1 Aerospace Applications

The effects of radiation on polymeric materials continue to be of importance for aerospace programs as the design lifetimes of satellites and other space vehicles are increased [21]. Advanced materials, such as graphite-fiber-reinforced composites for structural applications are desirable because

Table 7.2 Dose Required to Reduce Elongation at Break to 50% of Original, Under Low-Dose Rate Conditions in Air (a) Versus High-Dose Rate and/or Inert-Atmosphere (b) Conditions

Polymer Type	Required Dose (kGy)[a]	Required Dose (kGy)[b]
Phenol–formaldehyde (47% sawdust)	2×10^3	6×10^3
Polyester (15% glass fiber, 56% mineral filler)	2×10^3	5×10^4
Polystyrene	5×10^2	2×10^4
Chlorosulfonated PE ("Hypalon")	5×10^2	8×10^2
Polystyrene–polybutadiene blend	4×10^2	2×10^3
Poly(ethylene-*co*-vinyl acetate)	4×10^2	2×10^3
PET	3×10^2	3×10^3
EPM	3×10^2	7×10^2
CR ("Neoprene")	3×10^2	5×10^2
Silicone rubber	2×10^2	3×10^2
PVC plasticized	100	2×10^3
LDPE	100	9×10^2
Poly(ethylene-*co*-TFE)	80	6×10^2
NR	70	1.5×10^3
Polyamide, aliphatic	20	4×10^2
HDPE	15	3×10^2
PP	7	10^3
PTFE ("Teflon")	1	40

[a]*Irradiation at 5–50 Gy/h in air.*
[b]*Irradiation at very high dose rate (10^4 Gy/h or higher) and/or inert atmosphere.*

of their high strength to weight ratio. Such composites have to be sufficiently resistant to radiation to be useful for service as components of aerospace vehicles. Composite matrix materials for these applications include epoxy resins with radiation resistance at 10^3 kGy and the recently developed cyanate ester resins, which do not show change after exposure with 1 MeV electrons with a dose of 10^5 kGy [14].

Phenolics and polystyrene are sufficiently resistant up to 8×10^4 kGy. Other specialty polymers used in aerospace applications are PI, aromatic sulfones, and polyether ketones. Fluoroplastics (PVDF, FEP, PFA, and PTFE) are used in specialty wire insulation and exhibit sufficient resistance when irradiated in the absence of oxygen. PVDF is often used in a preirradiated form to impart stability [14]. Metal foil wrapping of wire and cables is an additional means of protection.

Elastomers are used mainly in gaskets, seals, and O-rings in highly shielded environments. Phenyl silicone rubber is capable of service up to 10^3 kGy, and fluorinated elastomers to 10 kGy; EPM exhibits service threshold values of 500 kGy [14].

7.2.2 Radiation Sterilization

γ-Ray, X-ray, and EB irradiation is being used increasingly for sterilization of medical and pharmaceutical items for convenience and, more importantly, because of concern about toxicity of chemical sterilants, namely ethylene oxide [22]. The radiation sterilization of biomedical polymeric materials, particularly implantable surgical devices, raises significant concerns. As pointed out earlier, polymers typically undergo some radiation-induced degradation leading to discoloration and associated deterioration in properties. Another problematic effect in some polymers that results from specific radiation-stabilizing chemistries is odor. The polymers that most commonly exhibit postirradiation odor are PE and PVC (rancid odor from certain oxidized plasticizers that are based on soybean or linseed oil). The odor can often be mitigated through the use of antioxidants, different processing temperatures, or selection of higher molecular weight polymer. Odor reduction can also be accomplished through the use of gas-permeable packaging (e.g., "Tyvek" or paper) and elevated temperature conditioning [23].

Thus plastic components ought to be sufficiently radiation tolerant, not discoloring or degrading under the exposures needed to eliminate bioburdens. The selection of materials for radiation sterilization should start with the following basic rules [23]:

- Use the highest molecular weight material (with the narrowest molecular weight distribution that is possible) for the application.
- Aromatic materials are more radiation resistant than are aliphatic materials.
- Amorphous materials are more radiation resistant than semicrystalline materials.

- Higher levels of antioxidants improve radiation resistance (in many cases, there is a limit beyond which there is no effect; the amount has to be determined).
- Low-density materials are more radiation resistant than high-density materials.
- Materials with small pendant groups are more radiation resistant.
- For semicrystalline materials, the lower the degree of crystallinity, the greater the radiation resistance.
- Materials with low oxygen permeability are more radiation resistant.

Most thermoplastics, essentially all thermosets, and most elastomers can withstand at least one irradiation sterilization (<50 kGy) without significant damage.

7.2.3 Nuclear Industry Applications

Polymeric components are widely used in nuclear power plants in equipment, which is important to the safety of the plant. Most polymeric materials can be used in a radiation environment of at least 109 kGy, and some can be used to 1013 kGy in limited applications [24]. High dose rates cause heating; some properties of plastics and elastomers are more or less temperature dependent. Moreover, the radiation-induced change in properties may depend on the temperature, the type of radiation, dose rate, molecular structure of the material, and mechanical stresses acting on the part in service.

In general, the following applications are common in the nuclear industry [24]:

- Gasket and seals
- Hoses, flexible tubing, and diaphragms
- Electrical insulations, wires, and cables
- Thermal insulation
- Potting and encapsulation compounds.

A variety of elastomeric seals are used for the nuclear reactors and auxiliary equipment, including [24]:

- Airlock seals
- Personnel hatch seals
- Torus and wet well access hatch seals
- Refueling canal hatch seals
- Airlock shaft seals

- Drywell head seals
- Inflatable door and valve seals.

Polymers used in the radiation areas of a nuclear fuel reprocessing plant include PE, PI, fluoroelastomers, and to a lesser degree poly(etheretherketone). They are used as seals and bearings if the radiation levels are low enough. For nonactive or low-active pipework, PTFE has long been used as a sealant. Elastomeric materials used in the fuel reprocessing plants are mainly based on NBR and EPDM [25].

Cables used in the nuclear industry have to be subjected to rigorous testing before they are used. Qualification tests (type-tests) of cables are designed to simulate in-service aging in order to predict conditions at the end of the service life [26,27]. Simulation of radiation-induced aging is part of the testing if the cables are used in nuclear parts or exposed to any kind of ionization radiation in service. When performing the test of the effects of radiation, it is important to take into consideration not only the absorbed dose but also the dose rate [28–32]. The IEEE 383 standard for type tests permits irradiation of the test samples at dose rates up to 10 kGy/h. However, experiments generally cannot yield predictive information in long-term applications characterized by low dose rates typically less than 1 kGy [33].

In a study [33], a series of tests was conducted on cables with different sheath/insulation combinations, namely PVC/PVC, PVC/PE, and XPE/XPE. Production (real) cables were irradiated at dose rates approximately 7, 30, and 100 Gy/h with the absorbed dose rising to 590 kGy. The elongation at break, the oxygen induction time (OIT), and density were measured. After the exposure, the measured values for both elongation and OIT showed induction periods at the dose rates of 30 and 100 kGy/h; during these induction periods the measured values were almost constant. The absorbed dose during the period for elongation at break amounted to 100–150 kGy and that for OIT 50 kGy. Density showed an induction period comparable to that for OIT. At the dose rate of 7 kGy/h no induction period was observed.

The measured characteristics of the XPE/XPE cable essentially did not change during the irradiation. It was concluded that the qualification tests for XPE/XPE cables at a dose rate of 100 Gy/h can give quite satisfactory results. For PVC/PE and PVC/PVC cables, a dose rate of less than 30 Gy/h seemed to be an acceptable compromise.

References

[1] Clough R. In: Mark HF, Kroschwitz JI, editors. Encyclopedia of polymer science and engineering, vol. 13. New York, NY: John Wiley & Sons; 1986. p. 672.

[2] Charlesby A. Atomic radiation and polymers. Oxford, UK: Pergamon Press; 1960.
[3] Parkinson W, Bopp C, Binder D, White J. J Phys Chem 1965;69:828.
[4] Kang H, Saito O, Dole M. J Am Chem Soc 1967;89:1980.
[5] Alexander P, Black R, Charlesby A. Proc R Soc London, Ser A 1965;232:31.
[6] Kettiar A. J Appl Polym Sci 1959;2:134.
[7] Jenkins RK. J Polym Sci Part A 1966;1:771.
[8] Lyons BJ, Lanza VL. In: Hawkins WL, editor. Polymer stabilization. New York, NY: John Wiley & Sons; 1972, [chapter 6].
[9] Timmerman R, Greyson W. J Appl Polym Sci 1962;6:456.
[10] Evans MB, Higgins GM, Turner DT. J Appl Polym Sci 1959;2:340.
[11] Golub MA. J Am Chem Soc 1958;80:1794.
[12] Chapiro A. Radiation chemistry of polymeric systems. New York, NY: John Wiley & Sons; 1962. p. 358.
[13] Miller AA. J Phys Chem 1959;63:1755.
[14] Willis PB. Survey of radiation effects on materials. Presentation at the OPFM instrument workshop; June 3, 2008.
[15] Sisman O, Bopp CD. Physical properties of irradiated plastics. USAEC report ORNL—928. Oak Ridge National Laboratory; June 29, 1951.
[16] Wilski H, Duch E, Leugering H, Rosinger S. Coll Poly Sci 1981; 259:818.
[17] Schonbacher H, Stolarz-Izycka A. Compilation of radiation damage test data, Part II: Thermosetting and thermoplastic resins. Report CERN 79–08, European Organization for Nuclear Research, Geneva; August 15, 1979.
[18] Seguchi T, Yamamoto Y, Seguchi T, Yamamoto Y, Yegya H. Hitachi Cable Rev 1985;4:37.
[19] Neudorfl P. Kolloid Polymere Zeitschrift 1965;204:38.
[20] Parkinson W, Binder D. Materials for nuclear applications. ASTM technical report # 276, 1960.
[21] Tenny DR, Slemp, WS. In: Reichmanis E, O'Donnell JH. editors. The effects of radiation on high-technology polymers. ACS symposium series 381. Washington, DC: American Chemical Society; 1989. p. 224.
[22] Clough R, Shalaby SW, editors. Radiation effects on polymers. ACS symposium series 475. Washington, DC: American Chemical Society; 1991.
[23] Hammerich KJ. Polymer materials selection for radiation-sterilized products. Medical device and diagnostic industry news products and suppliers; February 1, 2000.
[24] Ademar Benévolo Lugão. Selected polymers materials for nuclear applications. LAS/ANS 2010 symposium on "New technologies for the nuclear fuel cycle". Rio de Janeiro; May 22, 2010.

[25] Materials used in a nuclear fuel reprocessing plant. Mater World 1994; 2(12):628.
[26] Drobny JG. Polymers for electricity and electronics. Hoboken, NJ: John Wiley & Sons; 2012. p. 244.
[27] IAEA-TEC DOC-1188. International Atomic Energy Agency, Vienna, Austria; December 2000.
[28] IAEA-TEC DOC-511. International Atomic Energy Agency, Vienna, Austria; 1990.
[29] Gillen KT, Celina M, Clough RL. Radiat Chem Phys 1999;56:429.
[30] Wise J, Gillen. KT, Clough RL. Radiat Chem Phys 1997;49(5):565.
[31] Plaček V, Bartoníček B. Nucl Instrum Methods Phys Res Part B 2001;185:355.
[32] Matsui T, et al. Radiat Chem Phys 2002;63(2):193.
[33] Plaček V, Bartoníček B, Hnát V, Otáhal B. Nucl Instrum Methods Phys Res Part B 2003;208:448.

Recommended Further Reading

Cleland Mr, Galloway RA. Electron beam cross-linking of wire and cable insulation. Technical information series TIS-01812, IBA Industrial Inc., Edgewood, NY.
Drobny JG. Polymers for electricity and electronics. Hoboken, NJ: John Wiley & Sons; 2012.
Clough R, Shalaby SW, editors. Radiation effects on polymers. ACS symposium series 475. Washington, DC: American Chemical Society; 1991.
Tenny Dr, Slemp WS. In: Reichmanis E, O'Donnell JH. Editors. The effects of radiation on high-technology polymers. ACS symposium series 381. Washington, DC: American Chemical Society; 1989.
Clough R. In: Mark HF, Kroschwitz JI, editors. Encyclopedia of polymer science and engineering, vol. 13. New York, NY: John Wiley & Sons; 1986.
Schonbacher H, Stolarz-Izycka A. Compilation of radiation damage test data, Part II: Thermosetting and thermoplastic resins. Report CERN 79–08, European Organization for Nuclear Research, Geneva; August 15, 1979.
Lyons BJ, Lanza VL. In: Hawkins WL, editor. Polymer stabilization. New York, NY: John Wiley & Sons; 1972.
Chapiro A. Radiation chemistry of polymeric systems. New York, NY: John Wiley & Sons; 1962.
Charlesby A. Atomic radiation and polymers. Oxford, UK: Pergamon Press; 1960.
Parkinson W, Binder D. In: Materials for nuclear applications. ASTM Technical Report # 276 1960.

8 Dosimetry and Real-Time Process Monitoring

OUTLINE

Industrial radiation applications of ionizing radiation sources of photons (X- and γ-rays) and electrons encompass a wide range of absorbed doses. The measurement of the amount of radiant energy delivered by the source is important in order to determine if the equipment operates properly and to ensure that the product will be of required quality. Proven dosimetry systems are widely used in the development of new processes, validation, qualification, and quality control of established processes and archival documentation of day-to-day and plant-to-plant processing uniformity. Commonly, doses range about six orders of magnitude, dose rates about twelve orders of magnitude, and energies over two orders of magnitude [1].

8.1 Dosimetry

Industrial applications of ionizing radiation encompass a wide range of absorbed doses (about six orders of magnitude), dose rates (about twelve orders of magnitude), and energies (over two orders of magnitude) [1].

The absorbed dose is the quantity of interest to be measured in order to evaluate the effects of the radiation on the processed materials. The methods used for that are known as *dosimetry*. There is a multitude of established dosimetric methods and some of them are listed in Table 8.1.

The main functions of dosimetry are [1]:

- The dose measurement leading to successful processing with a sufficient uniformity.
- Process validation, including dose settings for meeting process specifications.

Drobny: Ionizing Radiation and Polymers.
DOI: http://dx.doi.org/10.1016/B978-1-4557-7881-2.00008-0

Table 8.1 Examples of Dosimetry Systems

Dosimeter Type	**Type of Readout**	**Examples**	**Absorbed Dose Range (Gy)**
Calorimeters	Temperature measurements	Graphite, water, polystyrene	10^1-10^4
Radiochromic films	Spectrophotometers	Dyed and clear plastic films	10^0-10^6
Inorganic crystals	EPR spectra	LiF, SiO_2	10^3-10^7
Organic crystals	EPR spectra	Alanine, sucrose, cellulose	10^0-10^5
Chemical solutions	Spectrophotometers	Ceric–cerous, organic acids	10^2-10^5
Semiconductors	Electrical measurements	Si diodes	10^0-10^4

- Process qualification, including the establishment and locations and values of minimum and maximum doses.
- Process verification, by the coordination of routine dosimetry and the monitoring of process parameters.
- Proper calibration and traceability to standards using reference dosimetry.

The choice of the method depends on the type of process. With EBs, the critical process parameters are beam energy, beam current, scanning factors and uniformity, beam pulse characteristics, and the configuration of the product being processed. The criteria for selecting suitable dosimetry are summarized below [1]:

- Calibrated response over a specified dose range for a given radiation type and energy
- Suitability of the given dosimetry system and its response over the dose range
- Reproducibility and stability of measurement
- Energy deposition characteristics (radiation type, spectrum, dose rate)
- Conditions before, during, and after irradiation
- Packaging, handling, geometric condition of radiation

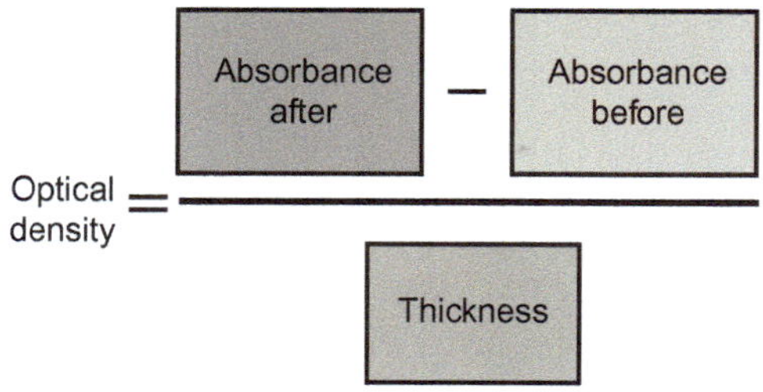

Figure 8.1 Equation for the Change in Optical Density of Film Dosimeters.

- Analytical technique and conditions and ease and speed of readout
- Dosimeter stability factors
- Cost and availability versus application.

The most convenient and widely used method is *film dosimetry*, which has been used for several decades. Commercially available film dosimeters are essentially thin plastic films either clear or containing dyes or dye precursors. These so-called *radiochromic films* are available in thickness ranging from 0.005 to 1 mm and are used to monitor EBs and γ-ray doses from 10 to 10^5 Gy [2]. These doses are typical for medical applications, radiation curing of coatings, adhesives, wire and cable insulation, cross-linking of shrinkable tubing and films and composites, etc. When irradiated, radiochromic films change irreversibly their optical absorbance in proportion to the absorbed dose. To minimize errors due to variation in the film thickness, the dosimeter response is usually expressed as the radiation-induced change in absorbance divided by the dosimeter film thickness (Figure 8.1). The radiochromic films are useful for measuring high-resolution absorbed-dose distributions of EB profiles and depth doses as well as dose distributions in thin layers such as on surfaces of irradiated materials. A partial list of currently available EB dosimeters is given in Table 8.2. The appropriate test procedures are described in several ASTM standards [3].[1]

There are several types of dosimeter systems that are in use for determining the absorbed dose. The simplest ones are *thin film dosimeters*, which are suitable for use in routine practice.

In general, readings of any dosimeter can be influenced by the following factors [4]:

[1]ASTM Standards E1204, E1261, E1275, and E1276 have been replaced by Standards ISO/ASTM51204, ISO/ASTM51261, ISO/ASTM51275, and ISO/ASTM51276, respectively.

Table 8.2 Currently Available Thin Film EB Dosimeters

Type	Film Thickness (μm)	Analysis Wavelength (nm)	Dose Range (kGy)
Radiochromic films with special dyes			
In polyamide (nylon)	50 or 10	600–510	0.5–200
In polychlorostyrene	50	630 or 430	1–300
In polyvinyl butyral	22	533	1–200
Radiochromic microcrystalline layer on a 100 μm polyester support	6 (sensor)	650 or 400	0.1–50
Cellulose triacetate	38 or 125	280	5–300
Dyed cellulose diacetate	130	390–450	10–500
Dyed (blue) Cellophane	20–30	650	5–300

- Dose rate or dose fractionation
- Temperature, relative humidity, oxygen content, light
- Its stability after irradiation.

Each type of dosimeter requires a specific procedure to ensure accurate and reproducible results, such as postirradiation heat treatment. Some need to be stabilized for a certain time (e.g., up to 24 h) before readings of absorbance are taken [5–7]. The absorbance reading can be done by conventional spectrophotometry or other more involved methods.

The original radiochromic films were blue Cellophane films colored by a blue disazo dye that could be bleached by ionizing radiation, with thickness between 19 and 26 μm, depending on manufacturer. Such thin transparent Cellophane films have been used as decorative heat-sealable and moisture-resistant packaging films. This might have been one of the reasons why this blue film was found not to be sufficiently reproducible in response to the radiation [8]. Improved Cellophane films have been produced specifically for dosimetry in two different intense colors, namely orange and violet with dissolved disazo dyes. They are delivered in thickness of 1 mil (25.6 μm) and maximum absorptions of 440 and 560 nm, respectively. Bleaching occurs gradually over the dose range 10–300 kGy [2]. Examples of radiochromic films are shown in Figure 8.2.

Figure 8.2 Examples of Radiochromic Dosimetry Films. (*Courtesy of Elektron Crosslinking AB.*)

Other commercially radiochromic films are PMMA, cellulose triacetate (CTA) [1,8], polyamide (Nylon) films, poly (vinyl butyrate) with pararosaniline and *p*-nitrobenzoic acid [1], and alanine films [2]. The CTA films are undyed; PMMA films are available undyed and dyed (red and amber) [2].

Thin film dosimeters are commonly used for the following measurements [4]:

- The *surface area rate* (or *processing coefficient*), relating to the area irradiated per unit time to the beam current and the absorbed dose, is determined by measuring the surface dose at several beam current levels.
- The *cure yield*, calculated in terms of the average absorbed dose delivered to the layer of interest multiplied by the conveyor speed per unit effective beam current, can be estimated by measuring the dose over the range of beam currents at different speeds.
- The *dose uniformity across the width of the beam* is measured by placing the film chips at about 1-inch (25 mm) intervals or using a long strip of film across the entire width of the beam.
- The depth–dose distribution is measured by irradiating a stack of radiochromic film chips with a thickness slightly greater than the practical range at the energy of interest. The depth dose is determined by evaluating the individual chips.

A dose reader designed for direct reading from exposed films offered by Elektron Crosslinking AB of Sweden is shown in Figure 8.3. A method for analysis of images of visibly colored dosimeter films is offered by a software

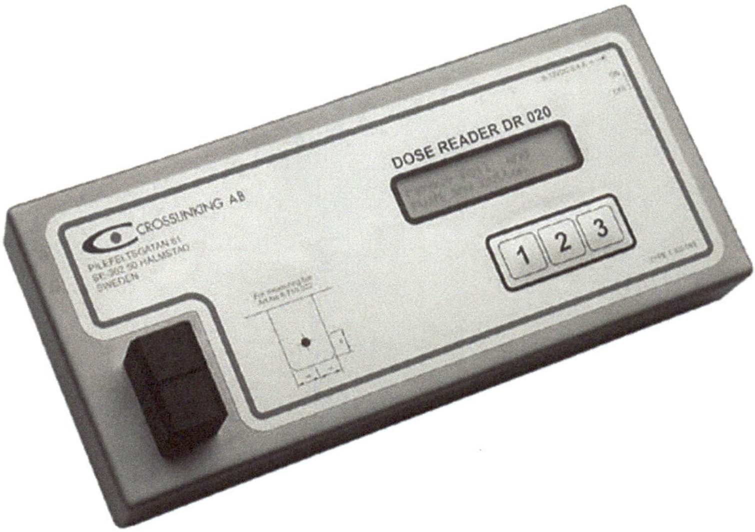

Figure 8.3 Dose Reader for Irradiated Films. (*Courtesy of Elektron Crosslinking AB.*)

package called RisøScan developed by Risø National Laboratory in Roskilde, Denmark. The colored dosimeter films can be scanned on an optical flatbed scanner and an image file created with information about the degree of coloration. A high spatial resolution of the scanned image can be obtained, but a compromise between the file size and resolution has to be obtained. Calibration is obtained by measurement of dosimeter films irradiated to known doses. Thus the software is a useful tool for measurement of dose and dose distribution using thin dosimeter films that color visibly [9].

To assure accuracy and reliable data, certain precautions must be taken when using radiochromic films [2,10]:

- They should not be irradiated in extremes of relative humidity, i.e., below 20% and above 80%. If such conditions cannot be avoided during irradiation, then the films should be sealed in PE pouches with the relative humidity between 50% and 70% before irradiation.
- Since the films are read optically, they should be kept free of dirt, scratches, and fingerprints.
- The films must be protected from UV irradiation at all times.
- In general, the films should not be read until approximately 24 h after irradiation, because full dye development requires several hours.

Figure 8.4 Graphite Calorimeter. (*Courtesy of Risø National Laboratory.*)

Any dosimeter used to determine absorbed dose in an irradiated product has to be calibrated [1,2]. The adiabatic character of EB deposition is used in calorimetry, which is the primary absolute method of measuring the absorbed dose (energy per unit mass) [11]. An example of the instrument used for this purpose is the water calorimeter developed in Risø National Laboratory in Denmark [12,13]. This calorimeter is reported to be suitable for electrons from a linear accelerator with energies higher than 5 MeV and shows accuracy $\pm 2\%$ [10]. Other types of instruments used for this purpose are the graphite calorimeter[14], shown in Figure 8.4 and the polystyrene calorimeter [15]. For example, a totally absorbing graphite body calorimeter is reported to be suitable for measuring energy deposited electron accelerators operating within the range 4–400 MeV [16–18]. Calorimeters made from polystyrene have been constructed for dose measurement for 1.5–10 MeV electron accelerators. Advantages of polystyrene as the absorbing material are the radiation stability of this material and the fact that identical phantoms can be made for irradiation of other dosimeters for calibration. This provides a precise tool for establishing traceability of dose measurements carried out at industrial electron accelerators [15].

Essentially, a calorimeter measures absorbed dose in terms of the radiation-induced temperature rise in the calorimeter core connected electrically to the outside. The dose to the dosimeter being calibrated is related to the associated calorimeter reading of the net temperature rise. These calorimeters are used as reference dosimeters for high-energy EBs (2–12 MeV) and also for small references and routine dosimeters (pellets and radiochromic films) [19].

The calibration is done by national or secondary standards laboratories. Calibration at National Institute of Standards and Technology (NIST) of the procedure involves exposure to either γ-ray source (^{60}Co) over the dose range 10^0-10^7 Gy or high-energy EB (1–28 MeV) combined with a graphite or a water calorimeter [1].

Alanine dosimeter operation is based on the ability of l-α alanine (a crystalline amino acid) to form a very stable free radical when subjected to ionizing radiation. The alanine-free radical yields an EPR (electron paramagnetic resonance) signal that is dose dependent, yet independent of the dose rate, energy type, and relatively insensitive to temperature and humidity [20]. Alanine dosimeters are available in the form of pellets or films and can be used for doses ranging from 10 Gy to 200 kGy [21]. A reference calibration service using alanine EPR system was developed and the scans were sent to the service center by mail. The currently available system allows transferring the EPR scan to a NIST server for calibration certificate. This way the procedure was shortened from days to hours [22].

Sunna dosimeter is a PE film containing LiF microcrystals, capable of measuring absorbed dose in a range of 50–300 kGy by measuring optically stimulated luminescence (OSL) (ASTM E2304-03) either in the visible or in the IR region of the spectrum. Another evaluation method, UV spectrophotometry, is also suitable for dose measurement in the 5–200 kGy range. The Sunna dosimeter has been found suitable for γ-, electron, and bremsstrahlung radiation by measuring the green OSL signal at 530 nm both below and above 1 kGy [23].

Ceric–cerous dosimetry is based on the potentiometric measurement of the reduction of ceric to cerous ions in an acidic solution by ionizing radiation [24]. The useful absorbed dose range for this system is from 500 Gy to 50 kGy. The dosimeters can be used as reference standard, transfer dosimeters, or routine dosimeters. This system is mainly used for dose determination in γ-ray irradiations with energies above 0.6 MeV, in X-ray (bremsstrahlung) irradiations with energies above 2 MeV, and in EB with energies greater than 8 MeV [24].

PE dosimetry is a method using PE films to indicate irradiating response [25]. Two ASTM International Standard Methods of Test (ASTM D6248 and ASTM F2381) use Fourier transform infrared spectroscopy (FTIR) to monitor the development of transvinylene (Figure 8.5) in irradiated PE (preferably higher crystallinity HDPE). The advantage of PE over other films is that it itself is very inert and can be made into films or molded test pieces of very precise dimensions. FTIR can also be used to indicate the degree of cure and cross-linking of acrylate monomers and oligomers in inks, coatings, and

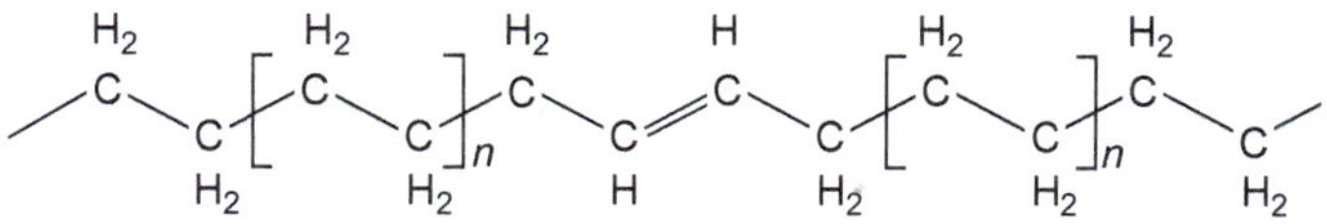

Figure 8.5 Transvinylene in Irradiated PE.

adhesives as well as in composite matrices [25]. More details on this subject are given in Ref. 25.

8.2 Real-Time Monitors

Real-time monitors are devices for the evaluation of ionizing radiation processes. They provide a continuous display in real time of dose delivered to product, as well as the energy of the EB. Moreover, they can show the variations in dose across a wide web if multiple detectors are installed and provide alarm signals to warn the operator of high- and low-dose conditions, and record the performance of the processor for production control, quality assurance, and maintenance needs. The major advantage of these real-time instruments is that they monitor the current and beam energy independently [26,27]. An energy monitoring device suitable for EBs with energy up to 7 MeV has been developed and evaluated in two industrial high-energy, high-power EB facilities. In both cases, the use of the energy monitoring device was found to assure a quick and reliable tool to control the actual value of the electron energy of the accelerator [28].

An electron dose monitoring and control system of low-energy electrons in EB radiation has been developed and is based on the measurement and calculations of the absorbed dose distribution of 0.6–2.0 MeV electrons in circular compound materials. From that a practical system for the absorbed dose control and monitoring of 0.3–3.0 MeV electrons in EB radiation processing of wires and cables was developed [29].

References

[1] McLaughlin WL, Desrosiers MF. Radiat Phys Chem 1995;46(4–6):1163.

[2] Humphreys JC, McLaughlin WL. IEEE Trans Nucl Sci 1981;NS-28 (2):1797.

[3] ASTM standards E170, E668, E1026, E1204, E1261, E1275, E1276.

[4] Mehnert R, Pincus A, Janorsky I, Stowe R, Berejka A. UV and EB technology and equipment, vol. 1. London/Chichester: SITA Technologies Ltd/John Wiley & Sons; 1998. p. 107.

[5] Janovsky I, Mehta K. Radiat Phys Chem 1994;43:407.
[6] Mc Laughlin WL, Puhl JM, Miller A. Radiat Phys Chem 1995;46:1227.
[7] Abdel-Fattah AA, Miller A. Radiat Phys Chem 1996;47:611.
[8] McLaughlin WL. Radiat Phys Chem 2003;67:561.
[9] Helt-Hansen J, Miller A. Radiat Phys Chem 2004;71:359.
[10] Farah K, Kuntz F, Kadri O, Ghedira L. Radiat Phys Chem 2004;71:337.
[11] Zagórsky Z. P. In: Singh A, Silverman J, editors. Radiation processing of polymers. Munich: Carl Hanser Verlag; 1992. p. 279.
[12] Brynjolfsson A, Thaarup G. Rep Riso, 53, Denmark: Risø National Laboratory; 1963.
[13] Holm NW, Berry RJ. Manual on radiation dosimetry. New York, NY: Marcel Dekker; 1970.
[14] McLaughlin WL, Walker ML, Humphreys JL. Radiat Phys Chem 1995;46(4–6):1235.
[15] Miller A, Kovacs A, Kuntz F. Radiat Phys Chem 1995;46(4–6, Pt. 2):1243.
[16] Mc Laughlin WL, et al. Dosimetry for radiation processing. London: Taylor & Francis; 1989.
[17] Sunaga H. et al. In: Proceedings of the sixth Japan–China bilateral symposium on radiation chemistry. Tokyo, Japan: Waseda University; 1994.
[18] Janovsky I, Miller A. Appl Radiat Isot 1987;38:931.
[19] Helt-Hansen J, et al. Radiat Phys Chem 2004;71:353.
[20] Garcia RMD, et al. Radiat Phys Chem 2004;71:375.
[21] Brochure "e-scan™ Alanine Dosimetry System". Bruker Biospin, GmbH, <www.bruker-biospin.com/> [accessed 22.07.12].
[22] Desrosiers MF, et al. Radiat Phys Chem 2004;71:373.
[23] Kovács A, et al. Radiat Phys Chem 2004;71:327.
[24] Dosimetric techniques, brochure from Kent State University. Middlefield, OH: NEO Beam Alliance Ltd, <http://ebeam@kent.edu/> [accessed 30.07.12].
[25] Industrial radiation processing with electron beams and X-rays. Vienna Austria: International Atomic Energy Agency; 2011. p. 52 <www.iaea.org>.
[26] Kneeland DR, Nablo SV, Weiss DE, Sinz TE. Radiat Phys Chem 1999;55:429.
[27] Korenev S, Korenev I, Rumega S, Grossman L. Radiat Phys Chem 2004;71:315.
[28] Fuochi PG, et al. Radiat Phys Chem 2009;78:481.
[29] Zhou X, Zhou Y, Zhou Y, Tang Q. Radiat Phys Chem 2002;63:267.

Recommended Further Reading

Industrial Radiation Processing with Electron Beams and X-rays. Vienna Austria: International Atomic Energy Agency; 2011 <www.iaea.org>.

Mehnert R, Pincus A, Janorsky I, Stowe R, Berejka A. UV and EB technology and equipment, vol. 1. London/Chichester: SITA Technologies Ltd/John Wiley & Sons; 1998.

Singh A, Silverman J, editors. Radiation processing of polymers. Munich: Carl Hanser Verlag; 1992.

Mc Laughlin WL, et al. Dosimetry for radiation processing. London: Taylor & Francis; 1989.

Holm NW, Berry RJ. Manual on radiation dosimetry. New York, NY: Marcel Dekker; 1970.

Charlesby A. Atomic radiation and polymers. Oxford: Pergamon Press; 1960.

9 Safety and Hygiene

OUTLINE

UV and ionizing radiation technology has been recognized as a process successfully using low or zero content of volatile organic compounds (VOC) for more than 25 years. This is a very important advantage over traditional processes in coating, printing, and adhesive applications using organic solvents and other volatile ingredients exclusively. However, as with any industrial manufacturing process, there are disadvantages and some hazards due to the equipment, procedures, and raw materials used. In UV/EB curing, these hazards are attributable to reactive chemicals, to some volatiles, and to the nature of the energy sources.

EB processors generate two types of *ionizing* radiation: the primary product is high-energy electrons, and a secondary product is, X-rays resulting from their interaction with matter. The ionizing radiation is damaging because of its capability of penetration into the human body. γ-Sources produce the damaging radiation directly and continuously. Therefore they require additional protective shielding.

The radiation curable liquid systems consist of a variety of chemical compounds, some of which may pose hazards due to their toxicity or tendency to irritate skin or eyes.

9.1 EB Equipment Health and Safety

As mentioned earlier, the stoppage of fast electrons generates X-rays, which are hazardous to human health. X-rays cause cell damage that can lead to cancer formation or genetic mutations. Even at small dosage, X-ray exposure may cause skin burning and general radiation syndrome.

Drobny: Ionizing Radiation and Polymers.
DOI: http://dx.doi.org/10.1016/B978-1-4557-7881-2.00009-2

Table 9.1 Maximum Permissible Dose Equivalents

Body Part	Maximum Dose Equivalent (rem)
Whole body, head, and trunk, active blood-forming organs or gonads	1.25
Hands and forearms	18.75
Skin on whole body	7.5

Manufacturers of EB equipment provide adequate shielding. Thickness of the shielding and the material used for it depends on the accelerating voltage. Typically, 1 in (25 mm) thick lead shielding is capable of stopping any X-rays generated by 300 keV accelerators. Most lower voltage EB processors are self-shielded, which means that the electron (and X-ray) source is completely enclosed by shielding. Removable parts of the shielding have to be equipped with safety interlocks so that the high voltage of the accelerator is turned off when the shielding is opened. EB units are provided with a radiation detector, which automatically shuts down the power if an alarm setting is exceeded [1]. Units with higher accelerating voltages, typically above 600 keV, require separate vaults constructed from concrete, steel, or their combinations enclosing the equipment. State and local governments in the USA have regulations controlling the use of radiation-producing equipment and all EB units must be licensed.

The current federal guidelines for occupational exposure to ionizing radiation in the United States are found in the Code of Federal Regulations, Part 1910.96. They stipulate that individual personnel working in the area where ionizing radiation is produced should not receive any period of a calendar quarter more than the dose equivalents listed in Table 9.1 [2].

Employees working in the area where EB equipment is operating must be monitored for exposure to ionizing (mainly X-ray) radiation by use of film badges, which detect and quantify any exposure to stray radiation. The employers must train workers thoroughly in the operation of the equipment and proper safety and hygiene, and keep records of exposure of their personnel to ionizing radiation. Occupational Safety and Health Administration (OSHA) has a standard that covers employee exposure to radiation: *Ionizing Radiation* (29 CFR 1910.1096) and copies of that standard are available from the web site http://www.osha.gov.

9.1.1 Personal Radiation Monitoring

Film badge dosimeters are used for monitoring cumulative exposure to ionizing radiation. The badge consists of a photographic film and a holder.

The film is sensitive and once developed, exposed areas increase in optical density (i.e., blacken) in response to incident radiation. The badge is typically worn on the outside clothing around the chest or torso.

Thermoluminiscent dosimeters (TLDs) measure ionizing radiation by measuring the amount of visible light emitted from a crystal in the detector when the crystal is heated. The two most common TLDs utilize calcium fluoride and lithium fluoride respectively. The former is used to record γ-exposure; the latter is for γ- and neutron exposure.

9.1.2 Radiation Monitoring Systems

Numerous specialized radiation monitoring systems are available, some of them being microprocessor-based for monitoring radiation by the use of detectors that are optimized for each application and type of radiation.

The RAD Alert radiation monitor has been developed by PCT Engineered Systems, a radiation area monitoring unit capable of detecting β-, γ-, and X-ray radiation. The monitoring station compares measured radiation levels against high and low reference levels. A signal loss alarm circuit ensures that the Geiger Müller sensor tube, high-voltage supply, and detection circuit are all operational. The system is designed for the use on industrial manufacturing equipment, such as EB processors. It is standard on all BroadBeam machines and is compatible with all brands of low-voltage EB equipment [3].

9.2 γ-Ray Health and Safety

Radiation processing with γ-sources poses an additional potential hazard to people (workers and general public) because large quantities of radioactive materials are located at one place for any purpose. For that reason, additional and more thorough protection measures are necessary. IAEA in cooperation with a number of international organizations issued basic safety standards (BSS) for protection against ionizing radiation and for the safety of radiation sources [4]. In 2003, the IAEA published a report that provides information and guidance regarding the design and safe operation of industrial irradiation facilities to organizations intending to purchase and operate them [5]. This information satisfies the requirements of the BSS in that a code is provided to ensure that during normal operation, maintenance, and decommissioning, and in emergency situations, radiation exposure of workers and public is within set limits and is kept *as low as reasonably achievable* (ALARA).

In 1992, the IAEA published a safety guide that provides device specific guidance regarding design, operation, and regulation of industrial irradiators [6]. Manufacturers of irradiators follow established procedures that satisfy national and international regulations regarding the design and manufacture of radiation sources, such as those in ISO Standard 2919 [7].

The operators and workers wear radiation dosimeters (badges) during working hours to monitor the amount of radiation dose they receive. These badges are read at regular intervals (e.g., every 2 weeks) to determine the dose received by the wearer of the badge. The IAEA, in collaboration with several international agencies, has set guidelines regarding safe limits of radiation dose that workers may receive [4], which are based on the recommendations of the International Commission on Radiological Protection (ICRP) [8].

9.3 Chemical Hazards

Materials used in EB curing present hazards mainly in contact with the skin and eyes and, to a lesser degree, accidental ingestion. In some operations, where spray equipment is used, there is some possibility of inhalation of volatilized materials.

Acrylate types of monomers and oligomers are known to be skin and eye irritants. Even if they do not cause irritation immediately, they may sensitize a person over a longer exposure and cause allergy [9]. On the other hand, they represent much lower hazard than common solvents (Table 9.2).

There are many aspects of protection of personnel when handling these chemicals in the production setting. First of all, they must be informed about the hazards involved by reading thoroughly Material Safety Data Sheets (MSDS) for each material used. These must be supplied by the manufacturer or vendor of the chemical and must be accessible to the personnel at all times. MSDS not only inform about the hazards, but also recommend proper handling, the type of personal protection to be used, ways of disposal, ways to handle spills, and emergency contacts.

Skin should be protected by gloves (preferably nitrile or butyl rubber), boots, and long-sleeve clothing. For the protection of eyes, safety glasses with side-shields or splash goggles are the best choice when handling liquid chemicals. If spray equipment is used, proper respiratory protection may be needed. If any contact with the chemicals occurs, the affected area has to washed with soap and water immediately; solvents should not be used. In extreme cases, proper medical help may be required. Adequate ventilation in the workplace is necessary. General and local exhaust systems should be installed to remove vapors and aerosols that may occur in operation. Respirators might be needed if engineering controls are not sufficient. If the product is a thin liquid at ambient or elevated temperature and may splash, a face shield might be necessary.

Table 9.2 Comparison of the Toxicity and Other Properties of EB-Curable Systems with Common Solvents

Chemical	Flash Point (°F)	VOC	Hazardous Waste	Systemic Skin Irritant	Toxicity	Reproductive Effects
TMPTA[a]	>212	No	No	Yes	No	No
Oligomers	$\gg 212$	No	No	Maybe	No	No
VM&P Naphtha	<0	Yes	Yes	Yes	Yes	No
Toluene	40	Yes	Yes	Yes	Yes	Yes
Xylene	100	Yes	Yes	Yes	Yes	Yes
1-Butanol	100	Yes	Yes	Yes	Yes	Yes

[a] *Trimethylolpropane triacrylate.*

In the European Union, a very comprehensive initiative with the aim to improve the protection of human health and the environment from the hazards of chemicals and to enhance the competitiveness of the EU chemicals industry is under way. It is referred to as REACH (EC 1907/2006), the acronym standing for Registration, Evaluation, Authorisation and Restriction of Chemical Substances. REACH places greater responsibility on industry to manage the risk of chemicals and provide appropriate safety information to professional users and, as far as the most hazardous substances are concerned, also to consumers. New substances have to be registered before they are placed on the market. REACH came into force on 1 June 2007, with a phased implementation over the next decade [10]. There is no doubt that it is markedly affecting the UV/EB industry.

References

[1] Golden R. RadTech Rep 1997;11(3):13.
[2] UV/EB curing primer 1. 4th ed. Northbrook, IL: RadTech International North America, vol. 57; 1995.
[3] RAD Alert Radiation Monitor, Model RAD3005. PCT Engineered Systems LLC, 2006.
[4] International basic safety standards for protection against ionizing radiation and for the safety of radiation sources, safety series no. 115. Vienna, Austria: International Atomic Energy Agency; 1996 <www.iaea.org>.
[5] Practice specific model regulations: radiation safety of non-medical irradiation facilities, IAEA-TECDOC-1367. Vienna, Austria: International Atomic Energy Agency; 2003 <www.iaea.org>.
[6] Radiation safety of gamma and electron irradiation facilities, safety series no. 107. Vienna, Austria: International Atomic Energy Agency; 1992 <www.iaea.org>.
[7] Radiation protection—sealed radioactive sources—general requirements and classification, ISO 2919. Geneva: International Organization for Standardization; 1998.
[8] Recommendations of the ICRP, publication no. 60. Oxford and New York: International Commission on Radiological Protection, Pergamon Press; 1991.
[9] Bean AJ, Cortese J. FLEXO 2000;25(7):37.
[10] European Commission, Joint Research Centre, Institute for Health and Consumer Protection; 2009.

Recommended Further Reading

Practice specific model regulations: radiation safety of non-medical irradiation facilities, IAEA-TECDOC-1367. Vienna, Austria: International Atomic Energy Agency; 2003.

Radiation protection—sealed radioactive sources—general requirements and classification. ISO 2919. Geneva: International Organization for Standardization; 1998.

International basic safety standards for protection against ionizing radiation and for the safety of radiation sources, safety series no. 115. Vienna, Austria: International Atomic Energy Agency; 1996.

UV/EB curing primer 1. 4th ed. Northbrook, IL: RadTech International North America; 1995.

Radiation safety of gamma and electron irradiation facilities, safety series no. 107. Vienna, Austria: International Atomic Energy Agency; 1992.

Recommendations of the ICRP, publication no. 60. Oxford and New York: International Commission on Radiological Protection, Pergamon Press, 1991.

10 Current State of Technology and Trends

OUTLINE

Our modern lifestyle increasingly demands more from polymers: higher service temperatures, better processing, improved physical and mechanical properties, and improved resistance to aging under severe conditions, but also to decompose after finishing their function as packaging, etc. Moreover, they should not have adverse effects on our environment. Evidently, all these requirements cannot be met by one or even few individual materials or their blends. There are thousands of polymeric substances with specific physical and mechanical properties, and these can be modified not only by compounding with fillers, plasticizers, and a great selection of additives, but also by a multitude of processing techniques.

One of the processing techniques is irradiation of polymers and polymeric systems by ionizing rays. As shown throughout this book, the radiation technology can deeply change the properties of polymers and polymeric systems, very often with considerably less energy and less impact on the environment than with the conventional processes.

The key industry drivers today are more than ever based on cost reduction, reduced energy consumption, and environmental issues. The main components of economy where radiation technology has been growing over the past several decades are:

- Automotive applications (automotive plastics and composites, wiring and cable insulation, paints, protective coatings, adhesives, and fuel cells)
- Electronics and electrical devices and appliances (wires and cables, shrinkable tubing, displays, conformal coatings, adhesives, magnetic media, optical fibers, and battery separators)

Drobny: Ionizing Radiation and Polymers.
DOI: http://dx.doi.org/10.1016/B978-1-4557-7881-2.00010-9

- Aerospace and military applications (wiring and cables, advanced composites, rocket components, repair materials, and markings)
- Medical and health care (coatings, adhesives, sensors and probes, prosthetics, biomedical products, irradiation procedures, hydrogels, sterilization of medical devices, and supplies)
- Housing and construction (wood structures, laminates, composites, plumbing, wiring and cables, composites, displays, and optical fibers)
- Consumer products (packaging-shrinkable films, labels, displays, CDs, and DVDs).

10.1 Current New Developments and Trends in Equipment and Chemistry

The main drivers for both equipment and chemistry over the past several years have been cost reduction, reduction of energy consumption, and environmental issues (sustainability, "green" raw materials, reduced volatile organic compounds (VOCs), reduced toxicity, lowering odor, reduced defects, and manufacturing scrap).

Companies are increasingly considering sustainability [1] in their business decisions.

Radiation technology offers several sustainability features when compared to thermal curing:

- Reduced usage of energy
- Safer workplace
- Reduced use of solvents or their complete elimination
- Reduced usage of fossil fuels and reduced greenhouse gas emissions
- Reduced or entirely eliminated pollution controls
- Reduced transportation costs, safety measures, and recovery systems due to reduced or eliminated consumption of solvents
- Recyclable inks, coatings, and product wastes.

These features have positive influence on performance and on economic returns in addition to resultant substantial environmental benefits.

10.1.1 New Developments and Trends in Equipment

Most of the recent developments in EB equipment at the time of this writing have focused on low-energy systems. These systems typically operate in the range of about 70–150 kV. Such equipment is well suited for curing of inks and coatings used in package printing applications and can also address a variety of nonprinting converting applications.

10.1.2 Ultralow-Energy EB Systems

New applications are demanding smaller and more efficient systems that can integrate directly into existing and new process lines. Equipment manufacturers are responding to these requests by developing innovative new designs. One of them is ultralow-energy EB systems with acceleration voltages in the range from 15 to 25 kV. They are designed to be installed directly into a vacuum coating chamber [1]. Applications for this type of equipment include curing coatings that are vacuum deposited or roll coated on a web within the vacuum chamber possibly with a combination with metallization at a vacuum typically 10^{-5} Torr. A depth/dose curve generated from Monte Carlo simulation for an EB unit operating at 15 kV in a 10^{-15} Torr vacuum shows that the electrons can penetrate coating weights up to 4 g/m^2, i.e., 4 μm for coatings with a density of 1.0 g/cm^3.

10.1.3 Low-Profile Systems

Advances in printing press technology targeted at the flexible package market have opened more opportunities for EB systems. An important package printing technology is the use of variable sleeves on web offset presses. This style of press requires the EB system to accept a low web entry height. The web height will also vary as the sleeve diameters change. The latest EB system designs accommodate these web-handling requirements and maintain a "side fire" orientation. The "side fire" orientation is preferred for the maintenance access that is required to perform a window foil change. The small size of this design also facilitates easier retrofits onto existing production lines starting at a press exit height of 24 in (600 mm) [1]. An example of this design is shown in Figure 10.1.

10.1.4 Integrated Shield Roll

This patented (US 8106369 to PCT Engineered Systems LLC) [2] design uses a temperature controlled roll to support the material, while the roll simultaneously serves as a functional portion of the required shielding of the EB processing equipment [1]. The roll is precisely fitted to the mating surfaces (Figure 10.2). The results are reduced size and amounts of materials used, minimization of the volume that must be inerted with nitrogen, and easy access for threading and cleaning. The exposed roll surface is also ideal for integrating other processes such as coating heads, extruders, and nip rolls.

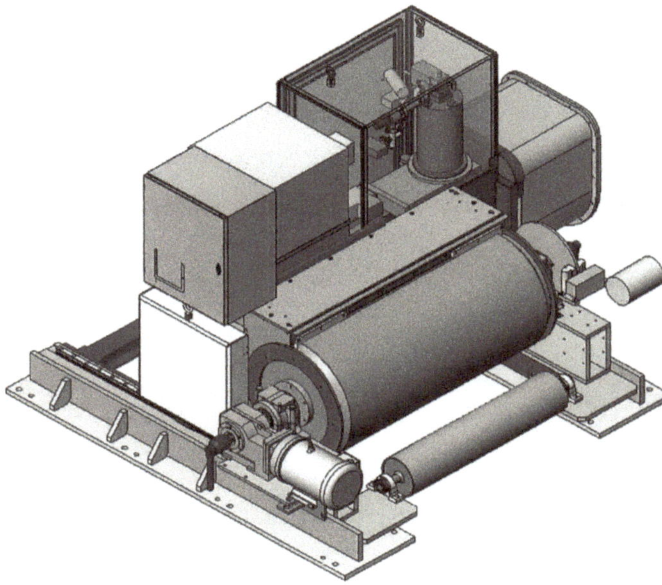

Figure 10.1 Schematic Representation of a Low-Profile EB System. *(Courtesy of PCT Engineering Systems, LLC.)*

Figure 10.2 Detail of the Integrated Shield Roll. *(Courtesy of PCT Engineering Systems, LLC.)*

10.1.5 EB Systems Based on Sealed Tube Emitters

Sealed tube EB emitters have been in the industry for several years. Recent advances in the reliability and configurations now allow these emitters to be applied in more industrial applications. Integration of these emitters into appropriate shielding configurations allows processing of a wide variety of materials including webs, flat materials, and three-dimensional objects. Recently developed sealed tube emitters are 16 in (400 mm) wide and allow processing at speeds up to 100 m/min (30 kGy dose) with voltages ranging from 90 to 180 kV [3,4]. An example of one version of this emitter is shown in Figure 10.3.

Web applications for EB systems based on sealed tube emitters may include narrow web printing, coating, and cross-linking. Potential applications for narrow web printing include:

- The curing of thick and/or high-density ink layers
- Supplemental curing of current UV inks
- Curing of EB inks and coatings designed for food packaging
- EB adhesive lamination
- EB cold foil transfer.

Some applications for this emitter are shown in Figure 10.4.

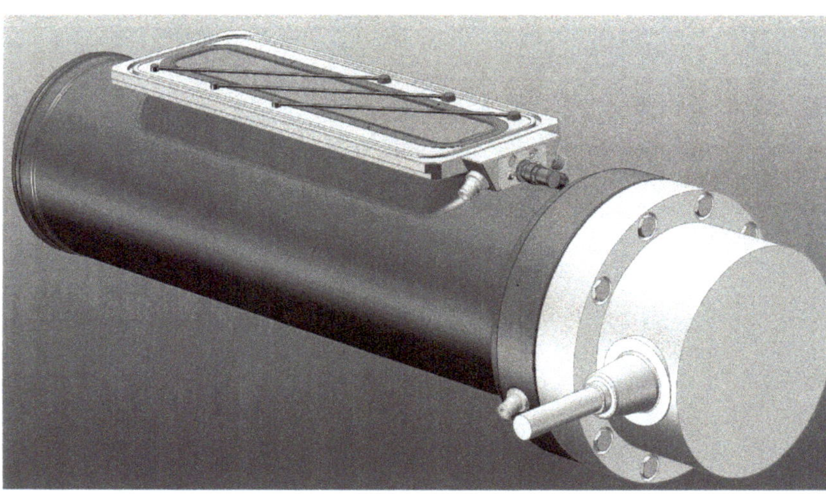

Figure 10.3 Comet e-beam Emitter EBA-200. *(Courtesy of Comet AG.)*

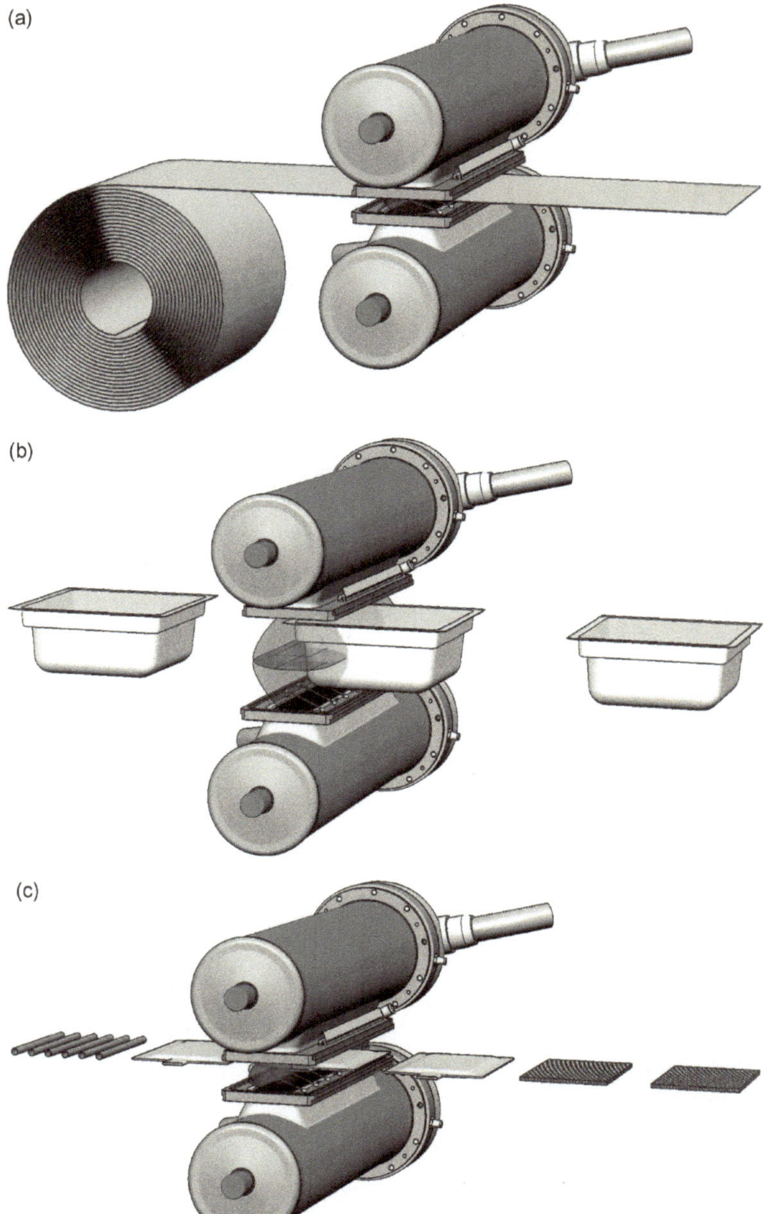

Figure 10.4 (a) 70–100 kV Sufficient for Irradiation of Flat Web—Curing of Ink or Sterilization on Surface. (b) 100–200 kV Sufficient for Irradiation of Nonflat Objects or Penetration of up to 200 μm. (c) 200–300 kV Allow Relatively Deep Penetration of up to 1 mm in Water (with Beam from Both Sides of the Web). *(Courtesy of Comet AG.)*

A narrow web system (see below) was developed that includes the above sealed tube emitter in combination with the integrated shield roll [1]. Another configuration of the seal tube emitter allows three-dimensional irradiation [1].

10.1.6 4 in 1 Flexible Packaging Line

The recently developed "4 in 1 System," is a line for flexible packaging that is comprised of four operations in a decorative coating press with the integrated shield roll [5]:

- Coating
- Lamination
- Cold foil transfer
- Cast and Cure™.

The patent-pending Cast and Cure™ system is an environmentally friendly decorative process that integrates casting and curing. It works on a variety of substrates and can produce ultrahigh gloss, matte, or holographic features. The system is depicted in Figure 10.5.

A major advantage of this system is that it can make packages more recyclable by eliminating the laminated metalized films used in the traditional holographic process. The film can be reused multiple times.

10.1.7 New Developments and Trends in Chemistry

1. The continuing efforts in chemistry relating to EB/UV cures has resulted in the considerable reduction or complete elimination of VOCs in coatings, adhesives, and inks, and using processes without them.

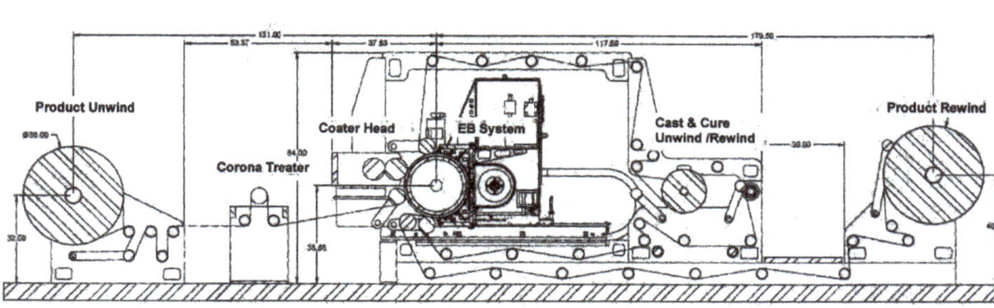

Figure 10.5 "4 in 1 System"—a Decorating Line with the Integrated Shield Roll. *(Courtesy of PCT Engineering Systems, LLC.)*

2. Research promoting sustainability had the goal of developing and using biorenewable oligomers for radiation cure, such as acrylated soybean oil, acrylated sucrose, and acrylated tetrahydrofurfural in coatings [6].
3. A published article deals with the concept of biorenewable carbon contents (%BRC). %BRC = number of biorenewable carbon atoms/number of total carbon atoms × 100 and introduces some examples of acrylated monomers and oligomers with BRC contents of 50% to almost 90% and functionalities of 2–12 acrylate groups per molecule [7].
4. A novel system has been developed for EB-initiated anionic polymerization using 4-isobutylphenyl-4′-methylphenyliodonium hexafluorophosphate (Irgacure 250 from BASF); cationic cure does not require nitrogen inerting. This system reduces equipment operating costs, since a continuous flow of nitrogen is not required [8].
5. *Atmospheric plasma processing* (APP) of film substrates under gaseous atmosphere leads to grafting of nitrogen-based chemical functions onto the surface of the substrates. APP-treated surfaces of polyolefin films, such as bias-oriented PP (BOPP), show excellent adhesion with EB/UV printing inks and clear coating thus eliminating the need for print-receptive primers [9].
6. Energy curable liquid systems for PSAs based on reactive acrylates represent an attractive alternative to traditional solvent-based adhesives from modified rubber. Besides their fast cure rate, they are environmentally more acceptable. The oligomers in these systems are a combination of monofunctional and difunctional acrylate oligomers based mainly on aliphatic urethane acrylates. They contribute to the cohesion of the bond. They have generally high molecular weight and low T_g to get the appropriate elastomeric behavior at room temperature. Properly chosen monomers act as reactive diluents and contribute to the viscosity of the liquid formulation, and to the cross-link density of the cured adhesive. Tackifying resins for the liquid PSA systems are most frequently C_5 or C_9 hydrocarbon resins. They usually have lower molecular weight and higher T_g than the oligomers [10].

10.2 New Developments and Trends in Process Technology

1. The printing industry and its importance are changing rapidly and a substantial part of it now uses electronic technology. It is only the package printing sector that is growing, which becoming the key

sustainable market for printing technology [11]. Economic and environmental pressures are greatly affecting package printing. The most obvious is the transition from solvent-based and water-borne inks and coatings to reactive media free of VOCs and water. Such inks and coatings can be "dried" only by radiation curing, i.e., by exposure to UV light or EB. Both these methods are very fast; UV cure has been successful in applications where the presence of photoinitiators and heat generated by the lamps do not matter. So UV cure cannot be used for food packaging where the migration of photoinitiators to food is not acceptable. Likewise some plastic films do not tolerate the heat from lamps. Thus in these cases, the remaining method of printing and decorating packaging materials is with EB-curable inks [12].

2. *Low-energy EB equipment* with acceleration voltages in the range 80–120 kV has been accepted during the past decade mainly in packaging. Gradually it has penetrated other markets. Currently, low-energy processors are used frequently for printing applications in the graphic arts industry where they are replacing UV curing units, and in web offset printing. In CI (central impression) flexographic printing, EB curing inks replace water-borne and solvent-based inks [12]. Because of the greater penetration of accelerated electrons of thick, multiple, and heavily pigmented layers of inks and coatings can be cured. EB curing enables wet-on-wet curing of several layers of inks at the end of the printing line with just one EB station. Another area where the low-energy EB systems are used successfully is the curing of overprint varnishes and high-gloss lacquers on gravure and flexo-printed materials and in the replacement of film laminates by EB-cured varnishes. In addition, low-energy systems can be used for curing coatings for decorative laminates [13].
3. *Fast check strips dosimeters* were evaluated for the determination of the dose delivered by low-energy processors in curing coatings, inks, laminated adhesives, and cross-linking of films. These strips are thin film strips coated with an EB-sensitive color change system. They are calibrated by using the Far West Technology (FWT) nylon radiochromic thin film dosimeter. After exposure, the strips are developed for 4–8 h and then read by β-color reader, which reads the color change of the dosimeter. They have been found to be useful for onsite evaluation of low-voltage EB accelerators with voltages in the range 80–125 kV. Results obtained in the study are within 5% of the results from FWT-60 radiochromic nylon dosimeters [14].
4. A *new thin film dosimeter* was developed that is based on triphenyl phosphene in a polyurethane matrix supported on 75 μm LDPE film.

The dosimeter is sensitive to EB irradiation levels below about 10 kGy. In addition, it is also sensitive to surface oxygen concentration and provides measure of inerting in the range, which affects curing of EB inks and coatings. Potentially this type of system can be used to assure good curing of EB inks and coatings [15].

10.3 Other New Development and Trends

10.3.1 Cellulose Degradation for Ethanol/Biofuel Production

Conversion of cellulose and sugars including sugarcane bagasse into ethanol or biofuel would avoid using corn and other agro-products that can also be used as food or animal feed. The conventional hydrolytic breakdown of these renewable resources generates contaminated water and undesirable toxins that interfere with fermentation. Pretreatment with EB facilitates the enzymatic conversion of these raw materials into alcohol [16–18].

10.3.2 Cellulose Degradation for Paper and Viscose Production

Large-scale testing of this process indicated that EB degradation of cellulose in wood chips would enhance the digestion rate for the production of paper or viscose. This promising technique has not yet been commercialized [16].

10.3.3 Recycling of Municipal Waste by Irradiation

A blend of recycled polymers consisting of 24% LDPE, 23% HDPE, 21% *cis*-PP, 15% PS, and 17% PET was irradiated by EB with doses up to 300 kGy with addition of 10% by weight of styrene–butadiene–styrene block copolymer grafted with maleic anhydride (SBS-*g*-MA) as compatibilizer. The SBS-*g*-MA contained 1.7% of MA and 30% of styrene. The effect of irradiation was evaluated by measuring Charpy impact strength and tensile impact strength of composites prepared from the above blend. The conclusion was that the composites with SBS-*g*-MA exhibit satisfactory impact strength values when irradiated with doses above 100 kGy and that the process is simple and inexpensive [19].

References

[1] Thompson TW. Advances in low-energy electron beam equipment technology. Paper presented at the RadTech UV&EB. Chicago, IL; April 30–May 2, 2012.
[2] Drenter J. U.S. patent 8106369 (March 10, 2012) to PCT Engineered Systems, LLC.
[3] Haag W. New generation electron beam emitter and laboratory unit. Paper presented at the RadTech Europe. Basel, Switzerland; October 18–20, 2011.
[4] Haag W. Sealed electron beam for use in narrow web curing, sterilization and laboratory applications. Paper presented at the RadTech UV&EB. Chicago, IL; April 30–May 2, 2012.
[5] Swanson KE. 4-in-1 Package decorating system: new EB converting equipment to enhance packaging materials. Paper presented at the RadTech UV&EB. Chicago, IL; April 30–May2, 2012.
[6] Chen Z, et al. RadTech Rep 2011;25(1):32.
[7] PCI (Paint & Coating Industry) Digital Edition; January, 2011, http://www.pcimag.com
[8] Lapin S. Electron beam-activated cationic curing. Paper presented at the RadTech UV&EB. Chicago, IL; April 30–May2, 2012.
[9] Cocolios P, Lapin S. Atmospheric plasma processing of film substrates for enhanced adhesion of EB curable inks and clear coats. Paper presented at RadTech Europe 11; October 18–20, 2011 [in Basel, Switzerland].
[10] Julian N, Cavalié H, Mélas M. Emphasis on UV/EB curable acrylate-based adhesives. Paper presented at the RadTech Europe 11; October 18–20, 2011 [in Basel, Switzerland].
[11] Laksin M. Electron beam in packaging: current and future trends. Paper presented at the RadTech UV&EB. Chicago, IL; April 30–May 2, 2012.
[12] Laksin M, Chatterjee S. U.S. patent 7997194 (August 16, 2011) to Ideon, LLC.
[13] Läuppi UV, Rangwalla I. Low-voltage electron beam curing—an update. Paper presented at the RadTech Europe 11; October 18–20, 2011 [in Basel, Switzerland].
[14] Grinewich O, Mejiritski A, Läuppi UV, Rangwalla I, Swain M. Quick and easy way to characterize low-voltage (60 to 125 kV) accelerators using fast check strips. Paper presented at the RadTech Europe 11; October 18–20, 2011 [in Basel, Switzerland].
[15] Lapin S. Oxygen sensitive electron beam dosimeters for measuring web surface inerting efficiency. RadTech Rep 2012;26(1):34.

[16] Industrial Radiation Processing with Electron Beams and X-rays. Vienna, Austria: International Atomic Energy Agency; May 1, 2011. p. 78.
[17] Duarte CL, et al. International meeting on radiation processing, IMRP 2008, London, UK; September 2008.
[18] Cardoso WM, et al. 8th International symposium on ionizing radiation and polymers. Brazil: Angra dos Reis, Rio de Janeiro; October 2008.
[19] Żienkewicz M, Dzwonkowski J. Polym Test 2007;26:903.

Appendix I
Major EB Equipment Manufacturers

Manufacturer	Web Page	Products
Budker Institute of Nuclear Physics (BINP)	http://www.inp.nsk.su/	Medium- to high-energy electron beam processors, X-ray irradiation equipment
Comet AG	www.comet-xray.com	Compact electron beam processors, laboratory units, X-ray equipment
Corex	corex.spb.ru	3–15 MeV electron accelerators with scanners
Elektron Crosslinking AB	www.crosslinking.com	Electron beam equipment, dose measuring equipment, laboratory unit
Energy Sciences Inc.	www.ebeam.com	Low- to medium-energy electron beam equipment, laboratory unit
IBA Industrial	www.iba-worldwide .com	Low- to high-electron beam and X-ray irradiation equipment
Mevex	www.mevex.com	Custom electron beam systems
NHV Corporation	www.nhv.jp	Electron beam and X-ray irradiation equipment, scanning and area beam designs
PCT Engineered Systems, LLC	www.teampct.com	Low- to medium-voltage electron beam equipment
Steigerwald Strahltechnik GmbH	www.steigerwald-eb.de	EB equipment mainly for welding, drilling, and research
Ushio America	www.ushio.com	Laboratory EB unit
Wasik Associates, Inc	www.wasik.com	Turn-key EB processing systems

Appendix II Standards and Specifications Applicable to Ionizing Radiation

AII.1 ASTM Standards

ASTM C637: Standard Specification for Aggregates for Radiation Shielding Concrete

ASTM C1321: Standard Practice for Installation and Use of Interior Radiation Control Costing Systems in Building Construction

ASTM D4082: Standard Test Method for Effects of Gamma Radiation on Coatings for Use in Nuclear Power Plants

ASTM D7767: Standard Test Method to Measure Volatiles from Radiation Curable Acrylate Monomers, Oligomers and Blends and Thin Coatings Made from Them

ASTM E170: Standard Terminology Relating to Radiation Measurements and Dosimetry

ASTM E668: Standard Practice for Application of Thermoluminescence-Dosimetry (TLD) Systems for Determining Absorbed Dose in Radiation-Hardness Testing of Electronic Devices

ASTM E1026: Standard Practice for Using the Fricke Reference-Standard Dosimetry System

ASTM E2232: Standard Guide for Selection and Use of Mathematical Methods for Calculating Absorbed Dose in Radiation Processing Applications

ASTM E2303: Standard Guide for Absorbed Dose Mapping in Radiation Processing Facilities

ASTM E2628: Practice of Dosimetry in Radiation Processing

ASTM D2701: Standard Guide for Performance Characterization of Dosimeters and Dosimeter Systems for Use in Radiation Processing

AII.2 ISO/ASTM Standards

ISO/ASTM 51204: Standard Practice for Dosimetry in Gamma Irradiation Facilities for Food Processing

ISO/ASTM 51261: Standard Guide for Selection and Calibration of Dosimetry Systems for Radiation Processing

ISO/ASTM 51275: Standard Practice for Use of Radiochromic Film Dosimetry System

ISO/ASTM 51276: Standard Practice for Use of Polymethylmethacrylate Dosimetry System

ISO/ASTM 51539: Standard Guide for Use of Radiation-Sensitive Indicators (in Radiation Processing)

ISO/ASTM 51702: Standard Practice for Dosimetry in a Gamma Irradiation Facility for Radiation Processing

ISO/ASTM 51707: Standard Guide for Estimating Uncertainties in Dosimetry for Radiation Processing

AII.3 Other Resources for Dosimetry

Guidance Notes for Dosimetric Aspects of Dose Setting Methods, The Panel on Gamma and Electron Irradiation, London, UK (1997)

Sharpe, P. and Miller, A., *Guidelines for the Calibration of Dosimeters for Use in Radiation Processing*, National Physical Laboratory, Teddington, UK (1999)

Appendix III Units of Exposure, Absorbed Dose, and Conversions

AIII.1 Exposure

The measure of ionizing ability of gamma rays is called the **exposure**.

- The *coulomb per kilogram* (C/kg) is the SI unit of exposure to ionizing radiation and is the amount of radiation required to create 1 coulomb of charge of each polarity in 1 kilogram of matter.
- The *röntgen* (R) is an obsolete traditional unit of exposure, which represented the amount of radiation required to create 1 esu of charge of each polarity in 1 cubic centimeter of dry air; 1 röntgen $= 2.58 \times 10^{-4}$ C/kg.

AIII.2 Absorbed Dose

However, the effect of gamma and other ionizing radiation on living tissue is more closely related to the amount of energy deposited rather than the charge. This is called the **absorbed dose**.

- The *gray* (Gy), which has units of (J/kg), is the SI unit of absorbed dose and is the amount of radiation required to deposit 1 joule of energy in 1 kilogram of any kind of matter.
- The *rad* is the (obsolete) corresponding traditional unit, equal to 0.01 joule deposited per kilogram; 100 rad = 1 Gy. A practical unit for absorbed dose is 1 kGy (1000 Gy).

AIII.3 Equivalent Dose

The **equivalent dose** is the measure of the biological effect of radiation on human tissue. For gamma rays it is equal to the absorbed dose.

- The *sievert* (Sv) is the SI unit of equivalent dose, which for gamma rays is numerically equal to the gray (Gy).
- The *rem* is the traditional unit of equivalent dose. For gamma rays, it is equal to the rad or 0.01 joule of energy deposited per 1 kilogram (0.01 Gy). 1 Sv = 100 rem.

Appendix IV Equation Used

1. The *Lambert–Beer Law* expresses the decrease of the intensity of radiation with the depth of penetration:

$$I_t = I_0 e^{-at}$$

where I_t is the intensity of the radiation after passing through the thickness t, I_0 is the initial intensity, and a is the coefficient of linear absorptivity.

2. The *Charlesby–Pinner equation* is used to determine the degrees of cross-linking and scissions from the soluble fraction of the irradiated polymer:

$$s + s^{1/2} = p_0/q_0 + 10/q_0 D u_1$$

where s is the soluble fraction, p_0 the density of scissions per unit dose, q_0 the density of cross-links per unit dose, D absorbed dose in kGy, and u_1 the number average molecular weight. Then the quantity p_0/q_0 represents the ratio of chain scissions to cross-links.

3. An empirical relationship expressing the energy of the electrons to their depth of penetration is due to Grun:

$$R_G = 4.57 E_0{}^{1.75}$$

where R_G is Grun range in μm and E_0 is the electron energy in keV.

The cross-link density can then be calculated from the degree of swelling using the following equations.

4. The *Flory–Rehner equation* is used to calculate cross-link density from the degree of swelling:

$$N = \frac{1}{2V_s} \frac{\ln(1 - \phi) + \phi + \chi\phi^2}{\phi^{1/3} - \phi/2}$$

where N is the moles of cross-links per unit volume (cross-link density), V_s is the molar volume of the solvent, ϕ is the volume fraction of polymer in the swollen gel, and χ is the polymer–solvent interaction parameter.

5. The *Mooney–Rivlin equation* is used for the calculation of the cross-link density from equilibrium stress–strain measurements:

$$\sigma/2(\lambda - \lambda^{-2}) = C_1 + C_2/\lambda$$

where σ is the engineering stress (force per unit of original cross-sectional area), λ is the extension ratio (measured length to original length), and C_1, C_2 = constants.

6. The *energy deposited* in the irradiated material causes a temperature rise (ΔT), which depends on absorbed dose and specific heat:

$$\Delta T = 0.239D/c,$$

where D is absorbed dose in kGy and c is specific heat in cal/g °C of the irradiated material.

7. The *cross-link density* ($1/M_c$) *of the product* is linearly proportional to the absorbed dose (D) and the initial prorad content (c_0):

$$1/M_c = (A + Kc_0)D$$

where A and K are constants; the constant K is dependent on the specific unsaturation.

8. *Throughput* (M/t) in mass/time:

$$M/t = 3.6 \times P \times f/D \text{ tons}/h$$

where P is the beam power (kW), D is the required dose (kGy), and f is the utilization efficiency.

9. *Delivered dose* = dose rate (Gy/s)/line speed.

10. *Delivered dose* = yield factor × beam current/line speed; it is used to characterize the curing performance of the electron processor. It is a constant that relates to the delivered dose, to the beam current and the line speed. The unit of yield factor is kGy × m/min/mA.
11. *Dose–speed capacity* = delivered dose × line speed

In usual EB operation, the dose–speed capacity is measured at 10 kGy. It is the most convenient unit to relate the curing performance of an electron beam processor to the desired process parameters, such as dose and line speed.

Appendix V
The 12 Principles of Green Chemistry

1. It is better to prevent waste than to treat or clean up waste after it is formed.
2. Synthetic methods should be designed to maximize the incorporation of all materials used in the process into the final product.
3. Whenever practicable, synthetic methodologies should be designed to use and generate substances that possess little or no toxicity to human health and the environment.
4. Chemical products should be designed to preserve efficacy of function while reducing toxicity.
5. The use of auxiliary substances (solvents, separation agents, etc.) should be made unnecessary whenever possible and such substances should be innocuous when employed.
6. Energy requirements should be recognized for their environmental and economic impacts and should be minimized. Synthetic methods should be conducted at ambient temperature and pressure.
7. A raw material feedstock should be renewable rather than depleting whenever technically and economically practicable.
8. Unnecessary derivatization (blocking group, protection, and deprotection, temporary modification of physical and chemical processes) should be avoided whenever possible.
9. Catalytic reagents (as selective as possible) are superior to stoichiometric reagents.
10. Chemical products should be designed so that at the end of their function they do not persist in the environment and break down into innocuous degradation products.
11. Analytical methodologies need to be further developed to allow for real-time, in-process monitoring and control prior to the formation of hazardous substances.
12. Substances and the form of a substance used in a chemical process should be chosen so as to minimize the potential for chemical accidents, including releases, explosions, and fires.

From: Anastas, P.T. and Warner, J.C. *Green Chemistry Theory and Practice*, Oxford University Press, New York, 1998 (with permission).

Bibliography

Below is a list of books which may provide additional and often detailed information on various topics in radiation science and technology. They are listed chronologically, starting with the most recent publications. Efforts were made to make this list as complete as possible.

Makuuchi K, Cheng S. Radiation processing of polymer materials and its industrial applications. Hoboken, NJ: John Wiley & Sons; 2012.

Industrial radiation processing with electron beams and X-rays. Vienna, Austria: International Atomic Energy Agency; 1 May 2011. <www.iaea.org>

Gamma irradiators for radiation processing. Vienna, Austria: International Atomic Energy Agency. <www.iaea.org>

Drobny JG. Radiation technology for polymers. 2nd ed. Boca Raton, FL: CRC Press; 2010.

Läuppi UV. EB/UV/γ—terms (English–German and German–English Dictionary). Hannover: Vincentz Network; 2003.

L'Annunziata M, Baradei M. Handbook of radioactivity analysis. Waltham, MA; Academic Press; 2003.

Koleske JV. Radiation curing of coatings. West Conshohocken, PA: ASTM International; 2002.

ASTM standards related to testing of radiation-cured coatings. Conshohocken, PA: ASTM International; 2001 [on CD-ROM].

Davidson RS. Exploring the science, technology and applications of UV and EB. London: SITA Technology Ltd; 1999.

Mehnert R, Pincus A, Janorsky I, Stowe R, Berejka A. UV & EB technology & equipment, vol. 1. Chichester, London, UK: John Wiley & Sons Ltd, SITA Technology Ltd; 1998.

Garrat PG. Strahlenhärtung. Hannover: Vincentz Verlag; 1996. p. 61 [In German].

Clough RL, Shalaby SW, editors. Irradiation of polymers, fundamentals and technological applications, ACS Symposium Series 620. Washington, DC: American Chemical Society; 1996.

Rechel C, editor. UV/EB curing primer: inks, coatings and adhesives. Northbrook, IL: RadTech International; 1995.

Mehnert R. Radiation chemistry: radiation induced polymerization. In: Ullmann's encyclopedia of industrial chemistry, vol. A22. VCH: Weinheim; 1993.

Singh A, Silverman J, editors. Radiation processing of polymers. Munich: Carl Hanser, Verlag; 1992.

Pappas SP, editor. Radiation curing science and technology. New York, NY: Plenum Press; 1992.

Clegg DW, Collyer AA, editors. Irradiation effects on polymers. London: Elsevier; 1991.

Clough RL, Shalaby SW, editors. Radiation effects on polymers, ACS Symposium Series 475. Washington, DC: American Chemical Society; 1991.

Farhataziz AM, Rodgers MAJ. Radiation chemistry. Weinheim: VCH; 1990.

McLaughlin WL, Chadwick KH, McDonald JC, Miller A. Dosimetry for radiation processing. London: Taylor & Francis; 1989.

Taniguchi N, Ikeda M, Miyamoto I, Miyazaki T. Energy-beam processing of materials. Oxford: Clarendon Press; 1989.

Bly JH. Electron beam processing. Yardley, PA: International Information Associates; 1988.

Randell DR, editor. Radiation curing of polymers. London: Royal Society of Chemistry; 1987.

Humphries S. Jr. Principles of charged particle acceleration. New York, NY: John Wiley & Sons; 1986.

Bradley R. Radiation technology handbook. New York, NY: Marcel Dekker; 1984.

Schiller S, Heisig U, Panzer S. Electron beam technology. London: John Wiley & Sons; 1983.

Kase KR, Nelson WR. Concepts of radiation chemistry. New York, NY: Pergamon Press; 1978.

Dole M, editor. Radiation chemistry of macromolecules. New York, NY: Academic Press; 1974.

Wilson JE. Radiation chemistry of monomers, polymers and plastics. New York, NY: Marcel Dekker; 1974.

Dole M, editor. Radiation chemistry of macromolecules, vols. 1 and 2. New York, NY: Academic Press; 1972, 1973.

Holm NW, Berry RJ. Manual on radiation dosimetry. New York, NY: Marcel Dekker; 1970.

Chapiro A. Radiation chemistry of polymeric systems. New York, NY: Wiley–Interscience; 1962.

Charlesby A. Atomic radiation and polymers. London: Academic Press; 1960.

Glossary

A

abatement (technology)—various processes and methods (e.g., incinerators) designed to eliminate or reduce the amount of hazardous waste, environmental emissions, or effluents from a facility.

abrasion—the surface loss of a material due to frictional forces.

abrasion resistance—reciprocal of abrasion loss.

absorbed dose—the amount of energy absorbed per unit mass of the irradiated material.

absorptivity (or absorption coefficient)—absorbance per unit thickness of the medium.

accelerator—1. equipment for the production of high-energy beams of elementary particles, such as electrons or protons through application of electrical and/or magnetic forces. 2. rubber compounding ingredient used in small amount with a vulcanizing agent to increase the rate of vulcanization (crosslinking) of the base elastomer(s).

addition polymerization—a type of polymerization in which the small molecules (monomers) combine chemically to form polymer molecules without a by-product material being formed.

adherend—a part covered by an adhesive and then joined to another. The surface to which an adhesive adheres.

adhesion—the state in which two surfaces are held together by interfacial forces, which may consist of valence forces or interlocking action or both.

adhesive—a material that when applied will cause two surfaces in contact with each other to adhere.

adhesive failure—failure of an adhesive bond at the adhesive–adherend interface.

adhesive strength—the strength of a bond formed by joining two materials using an adhesive.

aging—the irreversible change of material properties after exposure to an environment for an interval of time (also used as the term for exposing the material to an environment for an interval of time).

alpha particle (symbol α)—a positively charged particle emitted by certain radioactive materials; it is made up of two neutrons and two protons bound together, thus being identical to a helium atom, other than lacking the two electrons that He has. It is the least penetrating of the three most common types of radiation (alpha, beta, gamma) emitted by a radioactive material and can be stopped, for example, by a sheet of paper. It is not dangerous to plants, animals, or humans unless the alpha-emitting substance has entered the body.

ambient temperature—room temperature or temperature in the surrounding area.

amorphous phase—the part of a polymeric material that has no particular ordered arrangement in contrast to the crystalline phase, which is ordered. Semicrystalline polymers consist of different ratios of crystalline and amorphous phases.

amorphous polymer—a polymer having a noncrystalline or amorphous supramolecular structure or morphology. Amorphous polymers may have some molecular order but usually are substantially less ordered than crystalline polymers and subsequently have inferior mechanical properties.

angstrom (symbol Å)—a unit of length mainly used to express the wavelength of electromagnetic radiation and the size of atoms, and their components, equal to 10^{-8} cm (10^{-10} m). It is used very seldom, since it has been replaced by a new unit, the nanometer (nm); 1 nm = 10^{-9} m.

anionic polymerization—a process taking place by the addition of certain monomers to an active center bearing whole or partial negative charges.

annealing—a heat-treatment process directed at improving performance by removing stresses in the material formed during its fabrication. Typically, the part, sheet, or film is brought to a specific temperature for a definite period of time, and then cooled down slowly to ambient temperature.

antifoaming agent—a chemical that is added to a liquid mix to prevent formation of foam or added to the foam itself to break a foam already formed.

ASTM International—an international voluntary standards organization that develops and produces technical standards for materials, products, systems, and services. It developed from the original ASTM (acronym for American Society for Testing and Materials).

atomic mass unit—one-twelfth the mass of a neutral atom of the most abundant isotope of carbon, ^{12}C.

atomic number (symbol Z)—the number of protons in the nucleus of an atom and also its positive charge. Each chemical element has a characteristic atomic number.

atomic weight—the mass of an atom relative to other atoms. The atomic weight of an atom is approximately equal to the total number of protons and neutrons in its nucleus.

atomizing—dispersing solid or liquid particles into air.

B

back ionization—excessive build-up of charged particles limiting further deposition onto a substrate.

backscatter—reflected or scattered radiation in the general direction of the source after it strikes matter (gas, liquid, or solid).

bandwidth—the range of wavelengths between two identified limits.

bar—an SI unit of pressure equal to 1.0×10^5 pascals (Pa). It has the dimension of force per unit of area and is used to denote the pressure of gases, vapors, and liquids.

beam—a stream of particles or electromagnetic radiation traveling in a single direction.

becquerel (Bq)—SI unit of radioactivity, 1 dis/s.

beta particle (symbol β)—an elementary particle emitted from a nucleus during radioactive decay, with a single electrical charge and mass equal to 1/1837 that of a proton. A negatively charged beta particle is identical to an electron. A positively charged beta particle is called a positron. Beta radiation can cause burns, and beta-emitters are harmful if they enter the human body. Beta particles are stopped easily by a thin sheet of metal.

betatron—a doughnut-shaped accelerator in which electrons, traveling in an orbit of constant radius, are accelerated by a changing electromagnetic field. Energies up to 340 MeV have been attained in betatrons.

biaxial orientation—orientation in which the material is drawn in two directions, usually perpendicular to one another. Commonly used in films and sheets technology.

bioburden—the number of bacteria living on a surface that has not been sterilized or a measure of the contamination of an object with microorganisms.

Products or components used in the pharmaceutical or medical field require control of microbial levels during processing and handling by sterilization.

blowing agent—compounding ingredient used to produce gas by chemical or thermal action, or both, in the manufacture of hollow articles.

body ply—a calendered sheet consisting of one layer of rubber, one layer of reinforcing fabric (cord), and a second layer of rubber.

bremsstrahlung—electromagnetic radiation emitted by a fast moving charged particle (usually an electron), when it is slowed down (or accelerated) and deflected by the electric field surrounding a positively charged atomic nucleus. X-rays produced in ordinary X-ray machines are produced by this mechanism, as are X-rays produced during irradiation of materials by an electron beam. (In German, the term Bremsstrahlung means "braking radiation.")

butadiene—a gaseous hydrocarbon of the diolefin series. Can be polymerized into polybutadiene or copolymerized with styrene or acrylonitrile to produce SBR or NBR, respectively.

C

cable, electrical—either a stranded conductor with or without insulation, and other coverings (single-conductor cable) or a combination of conductors insulated from one another (multiple-conductor cable).

cable core—the portion of an insulated cable lying under the protective covering or coverings.

cable sheath—the protective covering applied to cables.

calender—a precision machine equipped with three or more heavy, internally heated or cooled (or both) rolls, revolving in opposite directions, with an adjustable gap between them. Calenders are used for continuously sheeting and plying up elastomeric compounds and frictioning or coating with elastomeric compounds and processing certain thermoplastic materials (e.g., PVC).

carbon black—finely divided carbon made by incomplete combustion or decomposition of natural gas or petroleum-based oils in different types of equipment. According to the process and raw material used, it can be furnace (e.g., HAF), thermal (e.g., MT), or channel carbon black (e.g., EPC), each having different characteristics, such as particle size, structure, and morphology. Addition of different types of carbon blacks to rubber compounds results in different processing behavior and vulcanizate properties.

cast film—a film produced by spreading or pouring resin dispersion, resin solution, or melt over a suitable temporary substrate (carrier), followed by solvent or water evaporation and/or melt cooling and removal from the substrate. Another method is to extrude a melt through a flat die on an extruder and cooling it on chilled rolls.

cathode rays—a stream of electrons emitted by a cathode, or negative electrode, of a gas-discharge tube or by a hot filament in a vacuum tube, such as a television tube.

cationic polymerization—a process in which the active end of the growing polymer is a positive ion.

chafer strip—a strip of rubber-coated fabric partially covering the bead assembly of a tire and extending above the rim line. Its function is the chafing of the bead on the rim.

channel black—a form of carbon black produced from natural gas by the channel process.

chlorosulfonated polyethylene—a product obtained by treatment of polyethylene by chlorine and sulfur dioxide. It is an elastomer highly resistant to chemicals and ozone; also known as "Hypalon".

cohesive failure—failure of an adhesive bond that occurs within the adhesive layer, leaving the adhesive attached to both adherends.

compound (elastomeric)—an intimate mixture of a polymer (or polymers) with all the ingredients necessary for the finished article.

composite—in polymer technology, a combination of a polymeric matrix and a reinforcing fiber with properties that the component materials do not have. The most common matrix resins are unsaturated thermosetting polyesters and epoxies and reinforcing fibers are glass, carbon, and aramid fibers. The reinforcing fibers may be continuous or discontinuous. Some matrix resins are thermoplastics.

compression molding—a fabrication method in which a polymeric material, mostly a thermoset (a plastic or an elastomer) is compressed in a heated mold for a specific period of time.

compression set—the residual deformation of a material after removal of the compressive stress.

contact angle—the angle a droplet or edge of a liquid forms with the solid plane of a substrate. A contact angle of zero degree indicates complete wetting; larger contact angle values represent limited wetting.

copolymer—a polymeric material formed by the reaction of two or more monomers.

cross-linking—a chemical reaction resulting in the formation of covalent bonds between chain-like polymeric molecules. As a result of cross-linking, polymers such as thermosetting resins become hard and infusible. Thermosetting (conventional) elastomers become stronger and more elastic; they do not dissolve in organic solvents, only swell.

crystalline melting point—the temperature or temperature range of melting the crystalline phase of a semicrystalline polymer. It is higher than the melting range of the surrounding amorphous phase.

crystallinity—orientation of the disordered long-chain molecules into repeated patterns. The degree of crystallinity (crystalline content) is expressed in weight percent.

crystallization temperature—temperature or temperature range at which a polymer crystallizes. The polymer chains which were randomly distributed in the melt become aligned into a close-packed ordered arrangement during the crystallization process.

cure (curing)—1. polymerization or cross-linking (or both) by radiation (UV or EB). 2. cross-linking or vulcanization of elastomeric materials.

curie (Ci)—traditional unit of radioactivity, $1\ \mathrm{Ci} = 3.7 \times 10^{10}\ \mathrm{Bq}$ (see becquerel).

cyclotron—a particle accelerator in which charged particles receive repeated synchronized accelerations by an electrical field as the particles spiral outward from their source. The particles are kept in the spiral by a powerful magnetic field.

D

damping—the dissipation of energy with time or distance.

degradation—deterioration, usually in the sense of a physical or chemical process, rather than a mechanical one.

dichroic—exhibiting significantly different reflection or transmission in two different wavelength ranges.

dielectric constant—the property of a dielectric (insulating material) which determines the electrostatic energy stored per unit volume for unit potential gradient.

dielectric loss angle (symbol δ)—an angle between the vector for the amplitude of the total current and that for the amplitude of the charging current. The tangent of this angle (tan δ) is the loss tangent, a direct measure of the dielectric loss.

dielectric strength—the voltage that an insulating material can withstand before breakdown occurs, usually expressed as voltage gradient (volts per mil or mm).

diffuse—a characteristic of a surface that reflects or scatters light in all directions.

dose—1. in EB processing: the amount of energy absorbed per unit mass, unit 1 gray (Gy) = 1 J/kg. 2. in UV processing: a common but loosely used term for *irradiant energy density* or *flux density* at a surface of the medium of interest (unit J/cm^2).

dose rate—the dose given in unit time and determined by the activity of the radiation source and the irradiation geometry. It is measured in, for example, kGy/h or Gy/s.

dosimeter—an instrument measuring absorbed dose.

durometer—an instrument used for measuring hardness of an elastomeric or plastic material.

durometer hardness—an arbitrary numerical value that indicates the resistance to indentation of the indentor point of a durometer.

dynamic properties—mechanical properties of polymeric materials exhibited under repeated cyclic deformation.

E

einstein—1 mole of photons.

elasticity—the property of a material by virtue of which it tends to return to its original size and shape after removal of the stress causing deformation such as stretching, compression, or torsion.

elastomer—a macromolecular (polymeric) material which, at room temperature, is capable of recovering substantially in shape and size after removal of a deforming force.

electromagnetic radiation—radiation consisting of associated and interacting electric and magnetic waves that travel at the speed of light (e.g., light, radio waves, gamma rays, X-rays); all can be transmitted through a vacuum.

electromagnetic spectrum—the full wavelength range of electromagnetic radiation, including microwave, ultraviolet, visible, and infrared energy.

electron—an elementary particle with a unit negative electrical charge and a mass 1/1837 that of a proton. Electrons surround the positively charged nucleus of an atom and determine the chemical properties of the atom.

electron beam radiation—ionizing radiation propagated by electrons accelerated by very high voltage (typically kilovolts to megavolts). This radiation is used frequently for cross-linking, polymerization, modifications, or degradation of polymeric materials.

electron volt (abbreviation eV)—the amount of kinetic energy gained by an electron when it is accelerated through an electrical potential difference of 1 volt. It is equivalent to 1.603×10^{-12} erg. Electron volt is a unit of energy, or work, not of voltage.

elongation—the extension of a body produced by a tensile stress.

elongation, ultimate—the elongation at the time of rupture under tensile loading.

emission spectrum—radiation from an atom in an excited state, usually displayed as radiant power versus wavelength. Each atom or molecule has a unique spectrum. The spectra can be observed as narrow line emissions (atomic emission spectra), or as quasi-continuous emissions (molecular emission spectra). A mercury plasma emits both line spectra and continuous spectra simultaneously.

energy density—amount of radiant energy arriving at a surface per unit of area, commonly expressed in J/cm^2 or mJ/cm^2. It is an integral of irradiance over time.

extrusion—a process in which heated or unheated polymeric material (plastic or elastomer) is forced through a shaping orifice (die) in one continuous shape, as in film, sheet, slab, profile, pipe, coating, etc.

F

fatigue, dynamic—the deterioration of a material by a repeated deformation.

filament—a small individual strand that is melt extruded during fiber spinning. Bundles of filaments are called fiber or yarns. Filament is also a fine wire (in a light bulb or electrical devices).

filament winding—a process to produce a composite part from a continuous fiber impregnated by a matrix resin by winding it onto a rotating or stationary mandrel.

fluence—the time integral of fluence rate (J/m^2 or J/cm^2). For a parallel and perpendicularly incident beam, not scattered or reflected, *energy density* and *fluence* become identical.

fluidization—the process of suspending powder particles using compressed air creating a fluid mixture of air and powder.

flux (radiant flux)—the flow of photons (in einstein/second).

free radical—a reactive species having an unpaired electron, which initiates a reaction with a double bond, for example, in acrylate polymerization. It is produced from its stable paired state by energy absorption. Scission of a covalent bond by thermal degradation can also produce a free radical.

free-radical polymerization—a process with a complex mechanism of initiation, propagation, and termination, of which the propagation and termination steps are typically very fast.

frequency—the number of times a periodic wavelength cycle occurs in 1 s; the unit of measurement is hertz (Hz).

frictioning—a process of impregnating woven fabric with rubber compound using a calender whose rolls rotate at different surface speeds.

furnace black—a carbon black obtained by incomplete burning of natural gas or petroleum oil, or both, in a large furnace.

G

gamma rays (symbol γ)—high energy, short-wavelength electromagnetic radiation. Gamma radiation frequently accompanies alpha and beta emissions and always accompanies nuclear fission. Gamma rays are very penetrating and are best stopped or shielded against by dense materials, such as lead or depleted uranium. Gamma rays are similar to X-rays but are usually more energetic and are nuclear in origin.

gel—a semisolid system consisting of a network of solid aggregates in which liquid is held.

glass transition (temperature), T_g—the temperature at which the amorphous portion of a semicrystalline solid changes from glassy state becoming soft and flexible (rubbery); not to be confused with melting temperature.

graft copolymer—a copolymer in which chains of one polymer are attached to chains of a previously formed polymer or copolymer in such a way that the junction points have three or more chains attached.

grafting—a reaction in which one or more species of block are connected to the main chain of a macromolecule as side chains having constitutional or configurational features that differ from those in the main chain.

gray (Gy)—an SI unit of radiation absorbed dose, equal to 1 J/kg; it has replaced the older unit *rad*, which is 100 ergs/g (or 0.01 J/kg).

graphite—a crystalline form of carbon. It occurs in nature but can also be produced by heating petroleum coke, carbon black, and organic materials. Used as lubricating filler or as an additive introducing electrical conductivity.

gravure coating—a coating technique using engraved rolls capable of picking up a precise amount of coating (disperse or ink) from a trough and transferring it to a web that is in contact with the roll. The engravings on the roll act as small reservoirs for the coating.

ground state—the state of nucleus, atom, molecule, or any other particle at its lowest (normal) energy level.

gum (compound)—an unfilled elastomeric compound containing only ingredients necessary for a sufficient cross-linking.

H

HAF (high abrasion furnace black)—highly reinforcing furnace carbon black increasing resistance of a rubber compound to abrasion.

half-life—a characteristic of a gamma ray source; time taken for the activity of a radiation source to decay to half of its original value.

heat aging—exposure of polymeric materials under specified conditions (temperature, time, presence or absence of air or oxygen, etc.), then testing them in stress–strain and hardness, determining the change of properties in comparison to the original (unaged materials).

hysteresis—energy loss, the difference between the work output in a cycle of extension and retraction.

hysteresis loop—in dynamic mechanical measurements, the closed curve representing successive stress–strain statuses of the material during cycle deformation; the area of the resulting loop is equal to the heat that is generated in the system.

I

impact resistance—the resistance to fracture under shock force.

injection molding—a molding procedure in which heat-softened thermoplastic material (melt) is forced from a cylinder into the mold cavity, which gives the article the desired shape. It is used for thermoplastics and for some thermosets.

inner liner—a layer of rubber with low permeability inside a tubeless tire, made of halobutyl rubber; its function is to assure that the tire will hold high-pressure air inside, without the air gradually diffusing through the rubber structure.

interpenetrating network (IPN)—a combination of two polymers into a stable interpenetrating network. In a true IPN each polymer is cross-linked to itself but not to the other, and the two polymers interpenetrate each other. In semi-IPN, only one of the polymers is cross-linked; the other is linear and by itself it would be a thermoplastic. The purpose of producing IPN is to improve strength, stiffness, and chemical resistance of certain polymeric systems.

ionization—the process of adding one or more electrons to, or removing one or more electrons from, atoms or molecules thereby creating ions. High temperatures, electrical discharges, nuclear radiation, or high-energy electrons can cause ionization.

ionizing radiation—any radiation displacing electrons from atoms or molecules (e.g., alpha, beta, and gamma radiations). Ionizing radiation may produce severe skin and tissue damage.

irradiation—the application of radiation to an object.

isotope—one or more atoms with the same atomic number (the same chemical element) but with different atomic weights. The nuclei of isotopes have the same number of protons but a different number of neutrons. Thus ^{12}C, ^{13}C, and ^{14}C are isotopes of the same element carbon, the superscripts denoting their differing mass numbers or approximate atomic weights. Isotopes usually have very nearly the same chemical properties, but somewhat different physical properties.

J

joule—a unit of work or energy (N m), abbreviated as J. It is a time integral of power.

K

klystron—a specialized linear-beam vacuum tube (evacuated electron tube). Klystrons are used as amplifiers at microwave and radio frequencies to produce low-power reference signals for superheterodyne radar receivers and to produce high-power carrier waves for applications such as radar and microwave relay transmitters, and are the driving force of modern particle accelerators.

L

laser—a device to produce an intense light beam with a narrow bandwidth. Laser is an acronym for *light amplification by stimulated emission of radiation.*

latex—an aqueous colloidal emulsion of an elastomer (natural or synthetic) or a plastic. It generally refers to the emulsion obtained from a tree or plant or the product of emulsion polymerization.

lethal dose—a dose of ionizing radiation sufficient to cause death. Median lethal dose (MLD or LD-50) is the dose required to kill within a specified period of time (usually 30 days) half of the individuals in a large group or organisms similarly exposed. The LD-50/30 for man is 400–500 rads (4–5 Gy).

linac—the short form for linear accelerator.

linear accelerator—a long straight tube (or series of tubes) in which charged particles (ordinarily electrons or protons) gain in energy by the action of oscillating electromagnetic fields.

linear energy transfer (LET)—the energy deposited along the track of ionizing particles.

line emission—the narrow lines of emission from an atom in the excited state, the "spikes" observed in spectrometry.

M

magnetron—the component contained inside a microwave-powered lamp head that converts high-voltage electrical input into microwave energy at 2450 MHz.

MDF (Medium Density Fiberboard)—a composite panel product manufactured from wood fibers and synthetic resin binders bonded together under heat and pressure; the fibers and resin form a homogeneous board with consistent properties in each direction.

melting point—the temperature at which the solid crystalline and liquid phases of a substance are in thermodynamic equilibrium. The melting point is usually referred to in the context of normal pressure of 1 atm.

melt processible polymer—a polymer that melts when heated to its melting point (or range) and forms a molten material with definite viscosity value at or somewhat above its melting temperature. Such a melt can be pumped and should flow when subjected to shear rate using commercial processing equipments such as extruders or molding machines.

micrometer (μm)—a unit of length equal to one millionth of a meter, replacing the older term *micron* that is still used occasionally.

microwave—designating the part of the electromagnetic spectrum associated with the larger infrared waves and the shorter radio waves, between 1 and 10 mm.

modulus—the ratio of stress to strain. In the physical testing of rubber, it is the force in pounds per square inch (psi) or pascals (Pa) related to the original area to produce a stated percentage elongation.

monochromatic—the light radiated from a source that is concentrated in only a very narrow wavelength range (bandwidth). This may be accomplished either by filters or by narrow-band emission.

monomer—a low molecular mass substance consisting of molecules capable of reacting with like and unlike molecules to form a polymer.

N

nanometer—a unit of length commonly used to define wavelength of light, particularly in the ultraviolet and visible ranges of the electromagnetic spectrum. It equals to 10^{-9} m or 10^{-3} μm or 10 Å.

neutron—an uncharged elementary particle with a mass slightly greater than that of the proton and found in the nucleus of every atom heavier than hydrogen (atomic weight = 1).

O

offset printing—a printing process in which the image to be printed is first applied to an intermediate carrier such as roll or plate, then being transferred to the substrate such as paper or film.

Oxygen (or oxidative) Induction Time (OIT) test—an accelerated aging test that is often used to predict the long-term stability of hydrocarbon materials including plastics, rubbers, and adhesives.

ozone (formula O_3)—an allotropic form of oxygen, produced by the action of electrical discharges or by that of ultraviolet light of certain wavelengths on oxygen. It is a gas with a characteristic odor and is a powerful oxidizing agent.

P

peak irradiance—the intense peak of focused power under a lamp, the maximum point of the irradiance profile. The unit of measurement is W/cm^2.

permanent set—the amount by which an elastic material fails to return to its original form after deformation.

photon—the carrier of a quantum of electromagnetic energy. Photons have an effective momentum but no mass or electrical charge.

phr—an abbreviation used in formulation of elastomeric compounds indicating parts in mass units per hundred parts of rubber.

ply—most commonly applies to a layer of rubber-coated fabric (as in tire or belt design).

polychromatic or polychromic—consisting of many wavelengths.

polymer—a macromolecular material formed by the chemical combination of monomers having either the same or different chemical composition.

polymerization—a chemical reaction in which the molecules of monomers are linked together to form large molecules whose molecular mass is multiple of that of the original substance. When two or more monomers are involved, the process is called copolymerization or heteropolymerization.

polyolefins—a large class of carbon chain elastomeric and thermoplastic polymers usually prepared by addition polymerization or copolymerization of olefins or alkenes, such as ethylene. The most important representatives of this group are polyethylene and polypropylene.

pressure-sensitive adhesive—an adhesive that requires applied pressure to the parts for the bonding to occur. It is usually composed from elastomer and modifying tackifiers. Pressure-sensitive adhesives are applied as melts or solvent-based systems. In most cases curing of the adhesive does not occur.

primer—a special reactive chemical dispersed in organic solvent or water applied to the substrate prior to the application of an adhesive or coating. A primer acts as a chemical bridge between the substrate and the adhesive or coating. Examples of primers are organosilanes and isocyanates.

proton—an elementary particle with a single positive electrical charge and a mass approximately 1837 times that of an electron, and is the nucleus of the ordinary or light hydrogen atom. Protons are constituents of all nuclei. The atomic number (Z) of an atom is equal to the number of protons in its nucleus.

Q

quantum yield—a measure of the photochemical efficiency of a photochemical reaction, expressed as the ratio of the number of chemical events per unit time to the number of photons absorbed per unit time; it is a unitless quantity.

R

rad (acronym for *radiation absorbed dose*)—the original basic unit of absorbed dose of ionizing radiation; 1 rad is equal to 100 ergs of radiation energy per gram of absorbing material. This unit has been replaced by the gray (Gy).

radiachromic—exhibiting a change of color or optical density with exposure to UV or EB radiation. These changes can be correlated to the amount of exposure.

radiant energy—energy transfer, expressed in joules or watt-seconds (1 J = 1 W s).

radiant power—rate of energy transfer, expressed in watts or joules per second (1 W = 1 J/s).

radiation dose (absorbed dose)—the amount of ionizing radiation absorbed by the unit of mass of the material during irradiation. The unit of radiation dose is a gray (Gy) defined as 1 joule per kilogram (J/kg). In practical applications a larger unit, namely kGy (10^3 Gy) is used. A previously used unit, no longer official since 1986, is the megarad (Mrad), equal to 10 kGy.

radiation illness—an acute organic disorder that follows exposure to relatively severe doses of ionizing radiation. It is characterized by nausea, vomiting, diarrhea, blood cell changes, and in later stages, hemorrhage and loss of hair.

radioactivity level—the strength (or power) of a radiation source, which is defined as the number of disintegrations of radioactive nuclides per second. The special name of the SI unit is becquerel (Bq). However, this is a very small amount of activity, and traditionally activity is measured in units of curie (Ci).

radiochromic—see *radiachromic*

radiometer—an instrument that senses irradiance incident on its sensor element and may incorporate either thermal or photonic detector.

rem (acronym for *roentgen equivalent man*)—the unit of dose of ionizing radiation, which produces the same biological effect as a unit of an absorbed dose of ordinary X-rays.

responsivity (spectral sensitivity)—the response or sensitivity of any system in terms of incident wavelength. In radiometry, it is the output of a device versus wavelength.

resilience—the ratio of energy output to energy input in a rapid (or instantaneous) full recovery of a deformed specimen.

RF (radio frequency)—any frequency between normally audible sound waves and the infrared portion of the spectrum lying between 10 kHz and 10^6 MHz.

RoHS (Restriction of Hazardous Substances)—restricts the use of certain dangerous substances in electrical and electronic devices.

rubber—a material that is capable of recovering from large deformations quickly and forcibly and can be, or already is, modified to a state in which it is essentially insoluble (but can swell) in boiling solvent such as benzene, methyl ethyl ketone, and ethanol–toluene azeotrope. A rubber in its modified state, free of diluents, retracts within 1 min to less than 1.5 times its original length after being stretched at room temperature to twice its length and held for 1 min before release.

S

SBS—styrene–butadiene–styrene block copolymer; a thermoplastic elastomer with polystyrene hard blocks and polybutadiene soft elastomeric midblock.

SEBS—styrene–ethylene–butylene–styrene block copolymer; a thermoplastic elastomer prepared by hydrogenation of SBS.

SEEPS—triblock styrenic copolymer with (ethylene–ethylene/propylene) elastomeric midblock; frequently used as compatibilizer.

SEPS—styrene–ethylene–propylene–styrene block copolymer; a thermoplastic elastomer. It is prepared by hydrogenation of SIS.

semicrystalline polymer—a material consisting of a combination of crystalline and amorphous regions. Essentially, all common plastics and elastomers with the tendency to crystallize are semicrystalline. The degree of crystallization depends on the structure of the polymer and the conditions of fabrication.

SIS—styrene–isoprene–styrene block copolymer with polystyrene hard segments and polyisoprene soft midblock.

SIBS—styrene–isoprene/butadiene–styrene block copolymer with polystyrene hard segments and a hybrid isoprene/butadiene soft midblock with controlled distribution of monomeric units.

solubility parameter—characterizes the capacity of a substance to be dissolved in another substance. It applies to solubility of polymers in solvents and to mixing of polymers (symbol is δ).

sulfur—chief vulcanizing agent for many elastomers, particularly those based entirely or partly on butadiene or isoprene.

T

tack (or building tack)—a property of an elastomer or rubber compounds that causes two layers of compound that have been pressed together to adhere firmly in the areas of contact. It is very important for building tires or other laminated structures.

target—material subjected to particle bombardment (as in an accelerator) or irradiation (as in a reactor) in order to induce nuclear reaction; the term is also used for the metal part used for the conversion of EB to X-rays.

tentering—a continuous process for drying cloth while it is held under tension to remove wrinkles and give a smooth surface; also for orienting polymeric films or sheets in the transverse (cross-machine) direction.

tentering frame—a machine used to stretch textiles and polymeric films and sheets.

terpolymer—a copolymer made from three different polymers.

thermal black—soft carbon black formed by the decomposition of natural gas (e.g., MT, medium thermal black). It has little stiffening effect but imparts toughness, resilience, good resistance to tearing, and fair abrasion resistance.

thermoplastic—a material capable of being repeatedly softened by increases of temperature and hardened by decreases of temperature (see plastics).

thermoplastic elastomer (TPE)—a class of copolymer or physical mix of polymers that is elastic at ambient or moderately elevated or lowered temperature and can be processed and recycled as a thermoplastic (i.e., by melt processing). The processing and service temperatures depend on the chemical nature of the material. (See also *TPE*.)

thixotropic liquid—a liquid that exhibits lower viscosity as shear rate increases. A practical example is house paint that becomes thinner when stirred.

tire, pneumatic—a tire casing (consisting of cord fabric) tread, sidewalls, and bead with or without an inner tube, capable of being inflated with a gas.

TPE—thermoplastic elastomer, consisting of a polymer or blend of polymers that has properties at its service temperature similar to those of vulcanized rubber but can be processed and reprocessed at elevated temperatures as a thermoplastic.

U

ultraviolet (UV) radiation—electromagnetic radiation in the 40–400 nm wavelength region. The Sun is the main natural source of UV radiation on the Earth. Artificial sources are many, including special UV lamps of different designs. UV radiation causes polymer degradation and other chemical reactions, including polymerization and cross-linking of monomeric and oligomeric systems.

V

van der Waals forces—weak attractive forces between molecules, weaker than hydrogen bonds and much weaker than covalent bonds.

viscosity—the internal resistance to flow exhibited by a fluid; the ratio of sheer stress to shear rate. A viscosity of one poise is equal to a force of one dyne per square centimeter that causes two parallel liquid surfaces of one square centimeter area and one centimeter apart to move past one another at a velocity of one centimeter per second. The SI unit of viscosity is pascal second (Pa s).

vulcanization—an irreversible process during which an elastomeric compound through a change in its chemical structure (e.g., cross-linking), becomes more elastic and more resistant to swelling by organic liquids, and the elastic properties are conferred, improved, or extended over a greater range of temperature.

W

watt—the unit of power equal to the work done at the rate of 1 joule per second or the power produced by a current of 1 ampere across a potential difference of 1 volt; 1/746 horsepower; abbreviated W, a smaller unit is a *milliwatt* (mW). In optics, it is a measure of radiant or irradiant power.

wavelength—a fundamental descriptor of electromagnetic energy, including light. It is the distance between corresponding points of a propagated wave; frequently used is the symbol λ (lambda). It is the velocity of light divided by equivalent frequency of oscillation associated with a photon. The unit of measurement is the *nanometer* (10^{-9} m).

Wehnelt cylinder—(also known as *Wehnelt cap*, *grid cap*) an electrode in the electron gun assembly of some thermionic devices used for focusing and control of the electron beam. It is named after Arthur Rudolph Berthold Wehnelt, a German physicist, who invented it during the years 1902 and 1903.

wetting—the spreading out (and sometimes absorption) of a fluid onto (or into) a surface. In adhesive bonding, wetting occurs when the surface tension of the liquid adhesive is lower than the critical tension of the surface being bonded. Good surface wetting is essential for high-strength adhesive bonding.

X

X-ray—a penetrating electromagnetic radiation emitted when the inner orbital electrons of an excited atom return to their normal state. X-rays are usually nonnuclear in origin and are generated by bombarding a metallic target with high-speed electrons.

Y

Young's modulus—in the elastic region, the relationship between stress and strain of a polymer undergoing tensile or compressive strain is linear (i.e., follows Hooke's Law). In this relation, the stress is proportional to strain. Young's modulus is the coefficient of proportionality in the stress–strain relationship.

Z

Z—the symbol for atomic number.

Index

Note: Page numbers followed by "*f*" and "*t*" refer to figures and tables, respectively.

F

R

FSC
www.fsc.org
MIX
Papier | Fördert
gute Waldnutzung
FSC® C083411

Kolibri 360 GmbH
Ferdinand-Jühlke-Straße 7
99095 Erfurt, Deutschland
produktsicherheit@kolibri360.de